Is Intelligence
an Algorithm?

Is Intelligence an Algorithm?

Antonin Tuynman PhD

BOOKS

Winchester, UK
Washington, USA

First published by iff Books, 2018
iff Books is an imprint of John Hunt Publishing Ltd., Laurel House, Station Approach,
Alresford, Hants, SO24 9JH, UK
office1@jhpbooks.net
www.johnhuntpublishing.com
www.iff-books.com

For distributor details and how to order please visit the 'Ordering' section on our website.

ISBN: 978 1 78535 670 4
978 1 78535 671 1 (ebook)
Library of Congress Control Number: 2017931224

A CIP catalogue record for this book is available from the British Library.

Design: Stuart Davies

Printed and bound by CPI Group (UK) Ltd, Croydon, CR0 4YY, UK

We operate a distinctive and ethical publishing philosophy in all
areas of our business, from our global network of authors to
production and worldwide distribution.

CONTENTS

Other books by Antonin Tuynman

Technovedanta:
Internet Architecture of a Quasiconscious Vedantic Webmind:
A Panpsychic Theory of Everything

Transcendental Metaphysics:
Technovedanta 2.0:
Transcendental Metaphysics of Pancomputational Panpsychism

Preface

The present book provides an analysis of intelligence. How do we understand the world around us? How do we solve problems? It has occurred to me that most often the answer to these questions follows a certain pattern, an algorithm if you wish. This is the case when our analytical left-brain side is at work. However, there are also elements in our behaviour where intelligence appears to follow a more elusive path, which cannot easily be pinned down to a specific sequence of steps. For instance our emotions and intuition cannot readily be described in an ontology of structural features and functions.

This book will, however, not only try to give you insight in intelligence as it functions in nature, like human or animal intelligence, but will also shed light on modern developments in the field of artificial intelligence. It also proposes further architectural solutions for the creation of a so-called global "Webmind". With respect to this aspect, this book continues where my previous book *Technovedanta*[1] had left open a number of then still unsolved problems.

Essentially I will try to draw a line and figure out where intelligence is algorithmic and where intelligence is what we could call "holistic".

The advantage of breaking down the analytical intelligence into understandable steps is that if we become conscious of these steps, we can also consciously implement them, avoid logical fallacies and develop a rapid strategy towards a solution, which does not need to evaluate all possible avenues. Such a strategy is called a heuristic.

Whereas many books have been written on the topic of devising heuristics, this has often been done in a specific context. Not in the least place by G. Pólya in his excellent book *How to Solve It*, which relates to mathematical problems. The present book provides general strategies for problems of any nature.

I hope that this attempt will be successful and provide you with a toolkit which allows faster pattern recognition and a more focussed strategy to find solutions to problems. What I ultimately hope to achieve is to increase your intelligence, as it gives you an enhanced freedom in dealing with the world around you and is a recipe for happiness.

Steemit

The way this book came about is a peculiar one, and needs a bit of an explanation. You can skip this part of the preface and jump to chapter 1 if you wish, because it is outside the framework of the topic of this book. But it may provide you with some insight into how I came to certain ideas explained herein.

In November 2016 I started writing a blog on the blog website "Steemit". Steemit is a very peculiar website. It is not just a platform to post articles and other creative products, no, it is a platform that rewards you for your contributions. Yes, you hear it well, you can be paid for your artistic concoctions.

You are paid in a currency called "Steem". This is a so-called cryptocurrency, like Bitcoin. This is not fake money, you can actually change Steem into Bitcoins and Bitcoins into Dollars via another platform called "Poloniex".

Where does this money come from? Cryptocurrencies are electronically "mined" in a computer network. Essentially, the computer is rewarded for solving difficult math problems. This processing power is used to verify transactions, so that number-crunching is required for the cryptocurrency to work. Users offer their computing power to verify and record cryptocurrency transactions into a so-called "blockchain".

According to Wikipedia a blockchain "is a distributed database that maintains a continuously-growing list of ordered records called *blocks*. Each block contains a timestamp and a link to a previous block. By design blockchains are inherently resistant to modification of the data – once recorded, the data in a block cannot be altered retroactively".

In Steemit, part of the blockchain is concerned with mining Steem and Steem-based transactions. Another important part is concerned with the posts submitted by the content creators.

Every day part of all the Steem mined by the miners is redistributed among the content creators.

Every post by a content creator is open for 24 hours for voting. If people have a lot of voting power, by voting they attribute value in Steem to a post. After the voting cycle is over, the major part of the Steem gathered is paid out to the content creator and a minor part is split among the people who voted for the post. Voting for a post is called curation and curators thus also receive a small reward.

People with an enormous amount of voting power are the so-called "Whales". If a whale votes for your post, suddenly the value goes up by the equivalent of about $10 or more. Active members with an intermediate amount of voting power are called Dolphins and can contribute the equivalent of a few cents to a few dollars of value.

The rest of the Steemit population are called the Minnows. They have not gathered enough voting power to attribute value. With every post you are rewarded for, your reputation index increases. The higher your reputation, the more likely people will vote for you.

You can convert your Steem into Steem Power, which is voting power. This transaction can also be reversed. You can also save your money in the form of Steem Dollars, which are a less volatile currency than Steem itself, which is prone to inflation.

The advantage of this system is that people are encouraged to provide interesting content and being rewarded stimulates them to continue. It is also an open system in the sense that the total amount of currency that can become available is not fixed. Somehow money is created out of thin air, but since there is trust in it, the system works wonderfully well.

It is here that I started my blog and since my series on "Is Intelligence an Algorithm" thrived and received quite some attention and rewards, I realised that this is information which people actually are interested in, and that's why I thought it useful to bundle it in a book. Of course you don't need to spend your money on this book, as you can read it entirely on the blockchain, but you will have to search for the different chapters.

Here you have them combined in a handy format with the bonus that I have completely referenced and cross-referenced it.

Qualifications

You may wonder what my qualifications are to write this book. I am neither a psychologist nor a neuroscientist. Instead I am a biochemist working as a patent examiner for the European Patent Office (EPO). It is, however, at the EPO where I became interested in artificial intelligence, heuristics and search engines. After all a patent examiner must search for prior art documents which have a similar or preferably the same content as a claimed invention. Moreover, a patent examiner must assess whether it was obvious to solve a given problem in the way proposed by the invention. To do so the EPO has developed a kind of algorithm, which is called the "problem solution-approach".

From all these notions I realised that what could actually be distilled as a pattern from this was a kind of algorithm for intelligence.

Just a few weeks after I had started writing, Dr Joe Tsien published an article (in *Frontiers in System Neuroscience*, Vol. 10, Article 95, 15 November 2016) showing that intelligence indeed follows a "neural network" type algorithm (not a traditional von Neumann style algorithm). I will address this in one of the last chapters of this book.

Nevertheless I will also try to show where and when intelligence does follow a sequence of steps which could be mimicked in a traditional von Neumann style algorithm, thereby going beyond the more "black box" approach of neural networks.

I will use analogies from various fields such as biology, psychology, chemistry, physics and even spirituality. I will also put in my own experience in abstracting and organising information into bite-sized chunks, which I have collected over the years and which allowed me to successfully pass a daunting exam like the European Qualifying Examination (EQE) for patent attorneys (which can also be sat by patent examiners, although this is not obligatory).

The EQE is worse than any test you may have had to pass at university. It comprises four tests which are very complex puzzles about inventions. You have to draft claims for an invention, you have to amend claims to overcome objections arising in the light of conflicting prior art, you have to prepare an opposition case (which is like a lawsuit) and advise a client how to deal with third party's rights that may prevent him/her from doing his business.

The exams last from 4 to 6 hours. Sometimes you end up with 30 densely written pages. Clearly you need information organisation skills here; clearly you need a heuristic.

Last but not least I have been an instructor at the patent academy for a couple of years and I have four children, each with a quite different learning curve. This has also prompted me to find ways to explain problems and solutions, to devise mnemonic and information organisation tools.

It is thus that I developed my "intelligence toolbox", which can be generalised to any type of field and problem and which I present here to you, so that you may also benefit from the tricks I discovered along the way.

I do not claim to provide you with an exhaustive treaty on what psychologists and neurologists think about intelligence. Rather I suggest useful pathways to organise information, solve problems and to creatively generate new information therefrom.

The book is accompanied by an appendix, which contains some further elements of chapter 1 as I originally wrote it. When I realised that I had started in a far too complex manner, I rewrote the chapter to logically fit in with the rest of the book. Since I do not wish to lose the richness of concepts I explored in this original chapter 1, I have included some of these as an appendix.

I hope you will enjoy my book.

Chapter 1

Nature's Meta-system transition algorithm for intelligence

If we follow the evolution of life on Earth, we see an increase in complexity of living systems, which has allowed life to adapt to very different environments.

A quite limited number of atoms such as carbon, hydrogen, oxygen, nitrogen, sulphur and phosphorous gave rise to simple organic molecules such as methane, ammonia, amino acids, phosphoric acid etc. These simple molecules formed the most basic building blocks to create more complex aggregates thereof in the form of macromolecules such as nucleic acids, proteins and lipids.

In turn the assembly of macromolecules in a complex architecture allowed for small prokaryotic organisms such as bacteria to evolve. With the advent of more complexity in the form of DNA and subcellular structures, the much bigger eukaryotic cells could form, which integrated prokaryotes as their engine to provide energy. Today we know these prokaryotes that still live inside our cells as the cell organelle which is called the mitochondrion.

Eukaryotic cells managed to cooperate and form yet more complex assemblies in the form of multicellular organisms. Whereas the first multicellular organisms were quite homogeneous as regards the types of cells they were built from, over time cells differentiated into different cell types and different cell types organised into organs.

Plants and animals arose in a versatile system of mutual exchange and mutual dependence. In the animal kingdom the limitations in size of the invertebrates, which were inherent to their heavy exoskeleton, were solved by inversing the support issue. The intelligent invention of an endoskeleton allowed for much bigger creatures to arise.

From fish to amphibians to reptiles to birds and mammals, we do not only see an increase in the abilities of the animals to adapt to their environment, we also see an increased tendency to be able to cooperate with the other members of the same species, giving rise to social structures, such as schools, flocks and herds.

In fact what we see as a pattern from these examples is that Nature evolved from simple singleton structures to more complex aggregated and integrated structures and that this is still an ongoing process.

Our human society is rapidly evolving to form a so-called "global brain". Not only do the transactions of goods and services establish a metaphorical global brain, with our invention of the Internet we are basically building a literal neural network which is sensing the whole planet via its Internet-of-Things sensor extensions.

We see that Nature has an inherent intelligence that appears to combine building blocks into bigger and more complex aggregates, which then in turn become new building blocks to build yet bigger and yet more complex aggregates.

This is in line with the definition which Ben Goertzel[2] the Godfather of Artificial (General) Intelligence gives for Intelligence:

The ability to achieve complex goals

Presently, we are living in the dawn of one of the greatest scientific breakthroughs: The very conceptualisation of the nature of intelligence, the self-organising pattern of the Universe.

In this chapter, I will discuss the way Nature appears to follow a kind of algorithmic pattern of what are called "meta-system transitions" when it is applying intelligence in order to evolve.

The aspects of human intelligence will be discussed in chapters 2–7, whereas chapters 8–10 are devoted to artificial intelligence. Chapter 11 discusses the elusive notion of "intuition" in Nature, humans and artificial systems.

In his book *Creating Internet Intelligence* Ben Goertzel[2] brings "Complexity science" to a higher level. Combining notions of Turchin's "metasystems transitions", Buddhism, General Systems and Network theory and Peircean and Palmerian metaphysics, he tries to define the very essence of Intelligence. The insights presented in this book are of such a profound nature, that they may well one day be recognised as the ultimate intelligence algorithm that underlies every phenomenon in this universe.

What the algorithm of Nature is constantly trying to achieve is to assemble existing parts to build new wholes such that the "whole is more than the sum of its parts". Or put in one word, Nature strives to achieve "Synergy" or "Emergence".

Nature appears to achieve this goal in a few basic steps:

Nature starts by providing elements which constitute pure "**Being**", which *prima facie* exist independent of anything else (e.g. atoms or subatomic particles). This is the provision of a dialectic thesis. The provision of such elements can be considered the first step of Nature's intelligence process.

A reaction to a stimulus from the environment shows that the elements do not exist alone but are relative to something else. This can be considered as the second step of Nature's algorithm, which is a reaction to a stimulus resulting in a "**Polarisation**". In dialectics this is the antithesis.

Now the different elements metaphorically aware of each other can establish a "**Relation**", for instance, the atoms start to form bonds between them as the third step of the algorithm. This is the process towards the dialectic synthesis.

From the web of relationships now a pattern can emerge, which forms the new aggregated entities. The constituent elements of these entities support and sustain each other so that their whole is greater than the sum of their parts. In the example a molecule is formed from the atoms. Its geometrical representation is a tetrahedron. This is the actual dialectic synthesis and the fourth step of the algorithm.

Ben Goertzel[2] has realised that this concept of "**Emergence**" is the key of evolution. This is how a mind's intelligence comes into existence: The combination of two or more parts can lead to a new phenomenon, a new entity in which the whole is more than the sum of the parts.

The new entity thus formed can be considered as a new "element" for building aggregates and can undergo this cycle again. This is how complexity arises in every system. It is the core of evolution and intelligence.

So the ontogenesis of holistic systems (i.e. systems where the whole is more than the sum of parts) is a four-step pattern or algorithm.

Now in my own words: 1) "Being" is followed by 2) "Polarisation or Reaction", the elements of which 3) engage in a plurality of "Relationships" from which 4) new entities "Emerge" by synergy.

Turchin's theories call the emergence of such a new meta-level a "**Metasystem transition**", which according to Goertzel amounts to the fourth step.

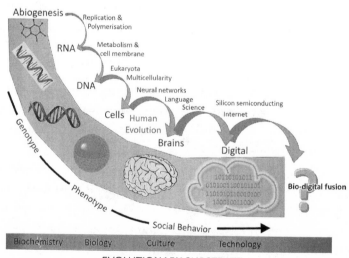

EVOLUTIONARY SUBSTRATE

based on: Gillings, M. R., Hilbert, M., & Kemp, D. J. (2016). Information in the Biosphere: Biological and Digital Worlds. *Trends in Ecology & Evolution*, 31(3), 180–189. http://escholarship.org/uc/item/38f4b791

Figure 1: Metasystem transitions: In the first transition molecules form macromolecules such as RNA and lipids. These form prokaryotic cells (e.g. bacteria) with a cell membrane in the second transition. Thirdly macromolecules recombine to form more complex macromolecules such as DNA from which emerge eukaryotic cells (e.g. yeast). From an ensemble of eukaryotic cells multicellular organisms can be formed in a fourth transition. Thus cells recombine into organisms (such as humans) with an internal architecture of differentiated cell types in different organs in what forms a fifth transition. In a sixth meta-system transition we are organising ourselves into a complex society, which now by virtue of the digital revolution is building a "Global Brain" in the form of the Internet and the associated Internet-of-Things (IoT) ecosystem. By Myworkforwiki – Own work, CC BY-SA 4.0, https://commons.wikimedia.org/w/index.php?curid=48996558.

Now I'd like to take Goertzel's concepts even further. If the "fourth step" results in a new entity as such, and therefore actually establishes a new instance of "Being", the step of a reaction to another entity on this aggregation level, a second entity, could be seen as a fifth step.

Note that in the Vedic tradition the so-called "5th chakra" is associated with creativity. Creativity requires inspiration, which in "Palmerian metaphysics" originates in so-called "Wild Being", arising from the interaction of what Palmer calls "hyper emergent entities": As a plethora of new different entities on a level starts to **recombine** among each other this creates diversity. This provides the element of the unexpected and establishes the non-deterministic stimulus for further development. The nature of this step doesn't appear to resemble a preprogrammed algorithmic step.

The relation that comes into existence in the process of creativity as the next step is the **distinction** of patterns, abstractions: a sixth step. Note that also the 6th chakra is associated with distinction.

From these trends then the new 7th level emerges, the sublimation and product of the creativity: A step of forming new entities, new knowledge and new intelligence as mental child: Athena born out of Zeus' head.

And thereby the circle of evolution on both microcosmic and macrocosmic level is round: The evolutionary process has in 7 steps returned to the essence of existence at yet a higher level of aggregation. "Seven" which is associated with Godhead in many cultural traditions: the 7 tones in music, the 7 colours of the rainbow.

The occurrence of "Sevens" in natural phenomena has even been suggested as being more than a coincidence as a consequence of the inner working of our brain according to R. Llinás[3] in the *I of the Vortex*. As the quantification constant of the so-called "Qualia" (the individual instances of subjective, conscious experience, such as colour, pain etc.); as a result of the Weber-Fechner law governing the intensity of sensory activation and perception ($s = klnA/A_0$); as organisational principle in biological systems (which gives rise to e.g. the geometrical structure of the shell curvature of the Mollusc Nautilus).

Note that Ben Goertzel[2] implicitly does mention these steps 5–7 as a repetition of steps 2–4 on a heterarchical level.

A key concept I cannot omit here is the fact that the patterns that emerge from a number of elements (which form at least a triad) can be expressed as "**Abstractions**": the expression of a simplification of the underlying phenomena. The pattern emerging from a triad a, b, c is the greatest common divisor at a different aggregation level. This pattern is a **representation** of phenomenal interactions **as something simpler**, which representation in itself is a new entity. This will become relevant at a later stage in this book.

It goes too far to discuss the mathematical and conceptual

framework of how Goertzel defines Mind, Meaning, Emergence, Attention Randomness, Complexity, Pattern etc. here, but I'm convinced he is on the right track to unravelling the mysteries of "Intelligence" as universal principle.

The unveiling of these steps shows that intelligence itself is a process type pattern; an **algorithm** that can be described and be put into practice.

Thus I have come to the conclusion that Nature's intelligence follows a basic algorithm, consisting of the seven above-mentioned steps.

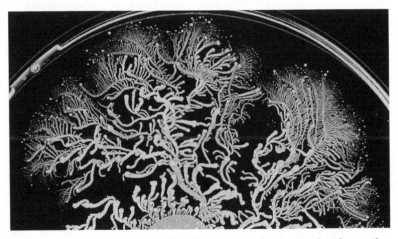

Figure 2: Colony of *Paenibacillus vortex* bacteria. Look at the uncanny resemblance to neuronal structures. Image by Prof Ben-Jacob, CC BY-SA 3.0, https://en.wikipedia.org/w/index.php?curid=33138215.

One thing that kept resonating in my mind was the congruence of this concept with the five elements of Ben-Jacob's[4] social learning machine (in bacterial colonies: the bacterial "creative web") about which I had read in the book *Global Brain* by Howard Bloom[5] (chapter "From Social Synapses to Social Ganglions"). This is also referred to as "Bacterial Wisdom".

Here is the parallel, which shows that the principles of

Peirce's, Palmer's and Goertzel's meta-system transitions follow in fact the same path of intelligence, the same algorithm that Nature also follows to evolve:

One: Bacterial colonies have a certain status quo in which a common language is imposed by a subgroup of bacteria called the "**conformity enforcers**" of the genome, to which all members chemically respond. This is the Thesis of the dialectic process before it is challenged; the first step. As long as there are enough resources in the environment to allow this thesis to persist, the colony can continue its population growth in a "boom" process.

Two: Colonies ultimately run into trouble as a consequence of exhausting their resources. This could lead to a "bust" or "fission" of the colony. Now a different subgroup of bacteria, the "**diversity generators**", is needed. These individual pioneers are needed to probe new alternative ways and resources: Mutants are generated which adapt to a changed scarcer environment and which are intended to become the discoverers of new resources. In any system this actually corresponds to the irritation or stimulus pointing to the incompleteness of the system: It creates the polarisation, the dialectic Antithesis. A second step.

Three: Enter the evaluation of the old paradigm vs. the new qualities of the pioneers, from within the species: The subgroup of bacteria called the "**inner judges**" enacts a comparator mechanism. Relations in the form of differences, correspondences and the spatiotemporal configuration of the new and old are probed and judged. A third step. In this **screen** those bacteria, which are failing outriders, commit suicide (elimination, **pruning**), whereas successful discoverers disperse an attractant of success. (In other systems, such as a computer system, the screening, pruning and determination of these relations would furthermore involve a classification and a ranking: hierarchical or heterarchical.)

Four: Depending on the circumstances either the old paradigm is maintained or if the mutants/discoverers are more successful, the resources are shifted towards the new heroes by a group of bacteria called the "**resource shifters**", thus establishing a new paradigm and new Thesis. The species has evolved due to the resource shifters.

This is the dialectic Synthesis: New features, which most often are the very distinguishing features between the old and new paradigm, have been added to the arsenal of the species and yielded an emergent property which gives an advantage over the species from which it originated. Goertzel's fourth step.

These are the four steps of intelligent evolution within the species. As Bloom[5] describes, these laws also function to create emergent Global Brains within higher social groups, such as beehives, anthills, but also among vertebrae, yes even among humans belonging to a group.

Above I went even further and added a fifth until a seventh step, which stages correspond to repeating steps 2–4 on a heterarchical level as described by Goertzel. Also these concepts really fit well in what Bloom describes as the fifth element:

Five: **Intergroup tournaments**. The newly established species with new emergent properties encounters other species with new emergent properties with which it will be in competition: This is the step where interaction of numerous emergent beings occurs, which Palmer calls "Wild Being". This is the new Antithesis. This competition between the species will have to lead to new diversity generation and creativity to overcome or join the other(s). This corresponds to the fifth step previously mentioned.

Six: Again a process of comparison occurs, which I will call "**distinction probing**", wherein the differences, correspondences and the spatiotemporal configuration of the new and old establish their relation and re-evaluate their strategies. This corresponds to the sixth step previously mentioned. This is again a process of "**screening and pruning**" in which a plethora of different solutions are tried out and only the best ones are kept, whereas the rest is eliminated. In Nature this is typically what happens in evolutionary living systems, with as best example the so-called Cambrian explosion.

Seven: That species which has an edge over the other due to superior distinguishing features or that species which advantageously can mimic or incorporate those features of the contender and add it to its own arsenal may come out of the battle as the victor. If this occurs a new synthesis has been arrived at. It has achieved a "fusion" of the characteristics of both contending species.

Imagine two primitive prehistoric human tribes, which did not know of their mutual existence, encountering each other. Pioneers will probe the strength of the other. If it is clear that the contending party has a serious advantage, the first party will withdraw and establish a niche elsewhere. If the strengths are deemed comparable it may come to a clash. Either one party has a superior advantage, which the other party is unable to incorporate, or one of the parties mimics that advantage successfully, so as to come to a strategy combining its own advantages with the advantages of the contender, thereby arriving at a meta-system transition with an even more superior advantage. Another scenario is that due to exchange of goods and habits in a peaceful way a new synthesis occurs.

There can be two types of synthesis: mere juxtaposition and true combination. If juxtaposition occurs, the old and new paradigms are of comparable strength, each having their own qualities and specialisations. This often leads to the formation of "Niches", wherein the contending species coexist.

A true combination or Symbiosis is a synthesis, wherein an exchange integral is present and both parties profit from each other in a win-win situation. Then real emergence has occurred and a new meta-system transition has been achieved. A good example in biology hereof is the symbiosis of bacteria and proto-eukaryotes from which the eukaryotes with their mitochondria emerged. This is again the phase of the establishment of a new paradigm and a meta-system transition giving rise to a new entity with new emergent properties.

Goertzel[2] was of the opinion that the division in 4 steps was in fact enough for his philosophy about existence. Steps 5–7 would merely be a repetition of steps 2–4 at the next aggregation level. The present seven-step scheme, which is more process oriented, describes evolution and intelligence as an algorithm, wherein the first four steps occur intra-species (within the species) as a reaction to a stimulus from the environment, which is at a lower aggregation level than the species itself. Steps 5–7 occur inter-species (between different species) and – in an ideal situation – lead to an exchange of features so as to give a new emergent entity. The devising of strategies and solutions internally is part of the first four steps. The learning from other entities is part of steps 5–7. In human intelligence, involving perception, cognition, reasoning and problem solving, similar steps are followed, as we'll see in the coming chapters.

Conclusion

This is Nature's intelligence algorithm to build aggregates of ever increasing complexity:

Nature provides a (living or non-living) system, an entity. (Thesis.)

The (living) system encounters a problem such as a lack of resources. This gives the system a **stimulus** to start to probe for a variety of alternatives or other solutions. (Antithesis.)

Nature will now generate a plethora of alternatives by combining elements from the environment with the system. (**Screening** of relational "Syntheses".)

From the probing or testing of these alternatives by the system, the system abstracts patterns. From these the most successful alternative strategies can be selected. (Elimination, **Pruning** of Syntheses and **Emergence** of new "Theses".)

This can be repeated on a heterarchical level between groups of entities or (living or non-living) systems (such as bacterial colonies or animal societies).

When contending groups encounter each other, this gives a **stimulus** to start a so-called "**Intergroup tournament**".

The tournament can lead to a mutual **probing of the distinctions** between the groups. Nature will **screen** which elements from the contender can be copied and **integrated** and which ones should be discarded.

This can result in the formation of 1) a "**niche**" (each group specialises in a niche such that it does not poach on the contender's preserves); 2) a "**symbiosis**": the groups learn to cohabitate peacefully together and provide each other with a service, resulting in a transactional scenario of a win-win situation; or 3) an exchange of those features which are different between the groups ("**mimicking**"). Thus the system adapts itself to its environment.

The most promising strategies ideally result in symbiosis, a unification of features toward which the system will strive. The system will try to resonate "morphogenetically" (i.e. in form, as dictated by its genetic make-up) with its new environment and thereby adapt to it. This is Nature's way of continuously striving for more complexity and incorporation of mutual features, as this assures more adaptability to and integration with the environment and hence increased chances for survival. In other words Nature's intelligence algorithm is essentially integrative: It tries to unite, to combine apparent opposites.

Note that such screening/pruning processes are not only happening in the realm of biological living systems; from the generation of a vast number of subatomic particle possibilities after the Big Bang, Nature selected a parsimonious number of only 57 particles, which make up the whole of existence! From these 57 only three, namely the proton, neutron and electron, essentially constitute the whole of the tangible material universe.

This is what Howard Bloom[5] calls Nature's evolutionary search engine, which employs fission-fusion strategies, or bust and boom cycles to create the beauty of complexity as we know it.

When I speak about Nature's intelligence algorithm, I do not mean that Nature implements a completely precooked routine. I do not mean that Nature is an automaton. Rather, what Nature implements is more like a "meta-algorithm": A sequence of steps that can be filled with recombined novel and inventive elements, so as to generate a specific algorithm to create specific types of complex structures. The variety of alternatives arrived at by combining elements is a creative process, as is the way Nature selects the winners.

The intelligence algorithm therefore goes beyond the traditional notion of an algorithm as a set of fixed instructions and is more like the modern "genetic algorithms" or "neural networks" used in artificial intelligence.

In the Appendix to this book you can read how this "algorithm" can be mimicked to be implemented in the creation of artificial intelligence.

In the next chapters I shall explore how Nature's intelligence algorithm is reflected in human intelligence.

Chapter 2

Cognition and Recognition

Many people are interested in improving their intelligence and many tools, schemes and tricks have been proposed for this purpose. Each of these tools (mnemotechniques, schemes to organise information, planning schemes, heuristics etc.) address only a small aspect of the complete process we call "intelligence", because most scholars dealing with the topic of intelligence do not have a good overview of the total picture of intelligence, what it is and how it functions.

If we could analyse intelligence and arrive at understanding its mechanism we might be able to use it to our advantage. This is the purpose of this book:

To collect a more complete understanding of what Intelligence is and how it functions, and to provide tools for improving our human intelligence as well as artificial intelligence.

In the previous chapter I already indicated that intelligence is a kind of algorithm, which I will only very briefly summarise.

When a living system encounters a problem such as a lack of resources it gets a stimulus to start to probe for a variety of alternatives or other solutions.

From its observations and testing it abstracts patterns. From these successful alternative strategies can be selected. When encountering contending groups so-called "Intergroup tournaments" can lead to a mutual probing of the distinctions between the groups. This can result in niching, a symbiosis, or an exchange of those features which are different between the groups. Thus the system adapts itself to its environment.

In this and coming chapters I will try to discuss different

aspects of Intelligence such as cognition, (pattern) recognition, memory, abstraction, analysis, understanding, information retrieval, reasoning and problem solving, including planning, heuristics and creativity, in more detail. I don't claim to present you with novel knowledge on this topic, but it is useful to give an overview of the teachings of the various Pundits in a simplified manner.

As said before, Ben Goertzel[2] the Godfather of Artificial (General) Intelligence defines intelligence as follows:

The ability to achieve complex goals.

This tells us what intelligence is about, its purpose, but it does not tell us how it functions. Since intelligence (at least analytical intelligence) appears to function in a certain predictable and repeatable way we could say it is a type of (natural) algorithm: A set of instructions defined in a very general way aiming to arrive at a goal, in this case a complex goal.

AN Whitehead[6] told us that "understanding is the apperception of pattern per se". Pattern recognition and understanding certainly are part of the way intelligence functions, but this does not tell us how understanding comes about and what to do with it once attained.

Another clue about the mechanics of intelligence is provided by Buckminster Fuller[7] who described cognitive processes as comprising four parts:

1 observation,
2 consideration (analysis),
3 understanding,
4 articulation.

This is a good starting point for analysing at least the first stages of the intelligence algorithm, which will be the topic of this chapter. In the next chapters I shall address the topics of reasoning, problem solving, planning and creativity.

When we observe an object or concept, we wish to know what it is, we wish to "cognise" it, so we analyse its form, its material, its constituent parts and the relations between those parts, its function, its purpose and how it relates to its environment.

This analysis is what B. Fuller calls "consideration": We build a network of relations, which gives us a framework for understanding, a constellation of facts or as B. Fuller calls it, a "consideration". Stella and Sidera are Latin words both meaning "star". A constellation or a consideration is a configuration of facts, metaphorical stars, which together describe a total form if you connect the "star dots".

From connecting the dots the framework that arises or emerges helps us to understand the object of analysis. The consideration framework metaphorically "stands under" the topic to be "understood".

Such a framework must geometrically speaking have at least three descriptive facts forming three relations, since with only two facts you only have one relation which does not build a plane for understanding. You need a third piece of information to identify what something is.

Imagine the information describing an object you get is the following: "black dots". This tells us nothing yet. If we get the information that there are only two of them at least we can start to speculate about its nature: Perhaps they are the plugholes of a socket, the nose of a pig, the double point symbol (colon) ":". Sometimes you need even more than three descriptors, if there is more than one thing fitting the information.

Additional pieces of information may suddenly tip the balancing point towards identification, such as a colour or a material nature of a substance. Once we have identified our observation by having a sufficient framework for understanding its "ontological configuration", we can articulate what we think it is.

We have now completed a cycle of the process of cognition or re-cognition.

When we re-cognise something we label the specific thing as an object belonging to a certain general pattern, a class or category. So we classify the object.

Every category has a certain group of features, which are typical for that category, and the presence of which in an object or concept form the requirements to belong to that category.

To describe an object or concept as complete as possible as regards its features in terms of material substance, form, internal relations of its parts, external relations with the environment, function, purpose, restrictions and rules etc. is the topic of "Ontology", the study of being.

Such a list characterising an object, phenomenon or concept is also called "an ontology".

To build a hierarchical classification of ontologies is building a kind of taxonomy, a classification scheme.

If we can picture a clear taxonomy in our mind, our recognition of objects, phenomena and concepts will dramatically improve. It allows for a rapid retrieval of information, namely to which type of pattern the object for consideration belongs: Our pattern recognition skills will improve.

We will have a clearer overview and distinction of which aspects are universal and extend across all classes, which aspects are general and belong to multiple classes and which aspects are specific for a class and thereby characterise it as a so-called "idiosyncrasy".

In fact our minds can form an ontology only if they have observed multiple instances of the same object. If you observe a new object for the first time, you can only try to make an approximation of what it is like. You can try to classify it in a higher ranking more general category if it fits in, and if it doesn't you can try to see which type of object is the most similar to it, i.e. which has a similar use and/or the most features in common.

Ben Goertzel[2] stated that "one is an instance, two a coincidence and three is a pattern". If we have observed three instances of a new object, which share the same features, it's worthwhile building an ontology for it.

When we build an ontology and recognise shared features in it, it means that from our memory we have been able to abstract aspects which the different instances of the phenomenon have in common.

Abstraction is a form of making a simplified generalised representation of something. For instance if we are able to abstract a tree to the structural features of single stem, branches, roots and leaves, it means that we have from all different types of stems been able to abstract the quality of "stemness", from all different types of leaves, the quality of "leaveness" etc. A simplified representation in words, which allows for recognising new trees as trees and to be able to distinguish them, for instance, from a bush.

Artificial General Intelligence is concerned with designing universal pattern recognition protocols, which inherently involve the process of abstraction. Abstraction always follows the pattern of going from multiple specific branches to a single generalised concept. In that way the process of abstraction itself could be represented by a tree-form.

Interestingly enough our neuronal structures also follow that pattern.

The ontology must also be linked to other ontologies, representing other objects, phenomena or concepts with which the one under investigation has a relation. The internal relations between the parts of the ontology must also be described as part of the ontology building.

You may have noticed that I make a distinction between objects, phenomena and concepts. Without wishing to define these fully at this moment, please note that with an object I mean a physical, tangible object, whereas with a phenomenon I wish not only to include objects but also non-tangible physical manifestations (e.g. light, sound). With a concept I wish to refer to mental representations or compounded ideas and schemes, which necessarily involve a degree of abstraction.

Thus far the first part of our intelligence protocol involves: observation, consideration – including pattern recognition and feature abstraction – and relation (web) building, giving a framework for understanding and thereby completing the building of an ontology, which subsequently allows us to articulate a (mental) abstract representation of the observation allowing for future recognition of further instances of the item of observation.

This is cognition, the transformation of raw percepts into appercepts. Or in JF Herbart's[8] terminology: "The process by which an aggregate or 'mass' of presentations becomes systematized (*apperceptions-system*) by the accretion of new elements."

From specific we have progressed to general, which makes it easier to store the item in our memory for future information retrieval and re-cognition.

This process can even be further improved by further abstracting the object to a simple "glyph". Chinese characters were formed as such simple glyphs. The Egyptians had their hieroglyphs. Alchemy used glyphs. Such simplified images representing most essential features are easier to retain mentally than complex lists of features in words.

In computers data are often compressed. When we have to learn long lists of information, we also have certain "mnemotechniques" at our disposal to compress this information. For instance we can form words or phrases built from the first letter of each most crucial term per item in the list. These words we can try to capture with an image thereof if possible or if not of a word that sounds similar and does have a visual representation. Ideally you form a set of glyphs, which you can mentally visualise as being arranged in a street through which you walk. Each house has a statue of the glyph in the garden at the number of the list.

These are great ways of compressing information and making them easily retrievable. Information retrieval can be improved by building webs of relations between the terms to be memorised so that together they form a single whole, a single configuration that can be glyphed or assembled in a kind of fairy tale.

Other mnemotechniques can be numerical coding using "Gematria"-type techniques in which words can be coded in numbers or vice versa numbers can be coded in words, depending on which mnemotechnique is more adapted for you. If you are a musician it can help to translate the number sequence into a melody, which you can remember, as each cipher from 1 to 10 can represent a note e.g. from C to E one octave higher. If you are artistically gifted you can try to put multiple visual representations in a picture as a whole that makes sense.

With these coding techniques we can improve the use of our storage space, storage speed, information retrieval speed and recognition speed.

These techniques can improve our learning abilities significantly. When it comes to memorising complex and huge quantities of information it is worthwhile to distil the most characterising word (or two words) of each phrase and reformulating the phrase in such a way that the characterising word (or words) is the answer to the question. Or you make a list of the characterising words and apply the first letter word building approach mentioned above.

It is very useful to identify, for each question you make, which type of question it is in terms of the so-called 6 Ws (Who, What, Why, Where, When and hoW). These questions are also typical for ontology building and can help you to remember an item more easily.

In this chapter I have described the first stages of the intelligence algorithm concerning observation, consideration, understanding and articulation in conjunction with mnemotechniques and ontology building tools to improve our abilities for these stages.

Chapter 3

Reasoning

One of the most important tools of the Intelligence algorithm is "Reasoning". Before we can explore the more challenging topics of "Problem Solving" and "Heuristics", which I will discuss in chapter 4 of this book, we must first get a thorough understanding of the process of reasoning, as we won't be able to devise Problem-Solving strategies without it.

In chapter 1, I have discussed the possibility that natural intelligence might be a kind of algorithm and what this can mean for the design of artificial general intelligence.

In chapter 2, I have discussed cognition, (pattern) recognition, memory, abstraction, analysis, understanding and information retrieval, as essential parts of the intelligence algorithm.

In this chapter I will first summarise what "reasoning" is and how it functions. Then I will try to show that it has an algorithmic nature and is in fact one of the integrated tools of intelligence. As part of this argumentation I will also touch upon the topics of rhetoric, causation, reality and truth.

Reasoning is often defined as thinking in a logical way to come to a judgement or conclusion. Logic itself is the set of rules we apply in this thinking process, it is the instrument used, but it is not identical to the thought process called "reasoning". The steps in reasoning to come from a premise (an assumption also called proposition which is believed to be true) to a conclusion we call "inference".

Traditionally there are three types of inferences: deduction, induction and abduction.

Deduction is the process in which a specific instance is compared with a general rule for a class of items assumed to be true. If the specific instance belongs to this class, it will also follow the rule.

E.g.

- All men are mortal (general rule for class).
- Socrates is a man (specific instance of class).
- Hence Socrates is mortal (conclusion).

Induction is the process in which multiple instances appear to follow a general pattern, from which it is then predicted that in the future further such an instance will follow the same pattern.

E.g.

- Until now the Sun always rose to start the day (general pattern).
- It is likely there will be a Tomorrow (specific future instance).
- I predict tomorrow the Sun will rise to start the day (predicted conclusion).

Abduction is the process in which for a specific instance (e.g. an observation, an effect) a reason or cause is speculated, which is known to give that result, whilst there can also be other causes yielding that result. It is another word for guessing.

E.g.

- The lawn is wet (specific instance which might or might not belong to a class).
- When it rains, the lawn is wet (general rule for class).
- Hence it has rained (conclusion).

Abduction is also called a logical fallacy, because the conclusion you arrive at need not be true; in the example given the lawn could also have been wetted by sprinklers.

Both induction and abduction are uncertain ways to come to

knowledge. Deduction is said to be the only certain process. But there is a snag here: The very premises of deduction (when they relate to physical phenomena) have been obtained by induction. In fact, we only assume that all men are mortal because until now we haven't seen an immortal one. Deduction works if the premises themselves are mental constructs the truth of which is asserted by definition, but that does not give us any certainty about the physical truth of such statements.

Induction and deduction, however, give us a strong probability that our conclusions will correspond to what can be observed.

There is also reasoning by analogy, which is also a logical fallacy. Because a specific instance belongs to a general class it is concluded that this specific instance has all the features of another instance of that class. This can lead to aberrant nonsense as illustrated hereunder:

- A man is a human.
- Beyoncé is human.
- Beyoncé is a man.

Oops...

A complete list with logical fallacies can be found on Wikipedia.

It is not my purpose in this chapter to treat each one of these in detail. But if you wish to increase your intelligence I recommend you have a look at these. It will improve your understanding of the world and people around you and you will be able to interact therewith and with them in a more intelligent manner. You will be able to avoid wrong conclusions that do not get you near the purpose of your intelligence, which – as I said before – is to achieve complex goals.

Logical fallacies often do not respect the principle that general rules or patterns must be **grounded** in a sufficient amount of independent observations (i.e. by different people). In order to arrive at a plausible conclusion, it must be probable. In order to be probable it must be statistically relevant and have a sufficiently large base of grounding observations.

Logical fallacies also arise due to an insufficient knowledge about general and specific and classification schemes.

One of the logical fallacies I do want to mention because of its significance in science is the correlative fallacy described by the Latin proverb: "post hoc, ergo propter hoc" (after this, therefore caused by this). If B comes after A, you conclude that A caused B. But we have plenty of examples in which correlation does not imply causation.

For instance, when people are eating more ice cream, there are more shark attacks. If you then conclude that eating ice cream causes shark attacks, you commit the above-mentioned fallacy.

Often correlated phenomena have a common underlying cause. In the case above, it was a warm day, which makes that more people swim in the sea, which attracts more sharks. The warm day also makes that people eat more ice cream.

How can we then assess what the cause of a phenomenon is? Scientists change one parameter, while keeping the others constant. If a change in parameter A systematically results in a more or less proportional change in effect B, they usually conclude that A causes B. This may be the case, and for the sake of being practical it is useful to assume it is, but it is not necessarily so.

When we reason on the basis of cause and effect, we are still stuck with a mechanistic type of thinking, which belongs to the seventeenth century. The universe of Newton, Huygens, Copernicus etc. in which the celestial bodies move with a clockwork precision, everything moving as if triggered by a plethora of cogwheels.

Quantum mechanics shows us that at the quantum levels

many of the deterministic premises do not hold. Since we are nothing but aggregates of quantum processes, how can we be so sure that cause and effect, as we believe they exist, really do exist? I will come back to this issue later in this chapter.

The branch of Artificial Intelligence (AI) that tried to work with parsing, specific rules and logical inferences has not been the most successful branch. It only leads to algorithms that can be applied in very specific contexts. Certainly useful in a specific context, but this will not out of itself evolve towards Artificial General Intelligence (AGI), which can operate independent of the context, let alone towards the human level of intelligence.

The more successful branch of AI, which works with Bayesian Networks and probabilistic inference, is much more based on correlations than cogwheel type cause-effect relations.

A good example thereof is "Latent Semantic Analysis" which is used in the IBM "DeepQA" engine Watson, known from an episode of the popular program *Jeopardy*, in which people competed with a computer in answering questions. Latent semantic analysis is based on Bayesian proximity co-occurrence of terminologies: If terminologies occur together in a statistical relevant way, they belong together and together they provide context and meaning.

In fact this is possibly also the way the brain builds ontologies when it is abstracting patterns of features and relations. Every ontology is said to be at least a "didensity": You need at least two terminologies to arrive at a relation, which provides meaning; a concept may even require three terminologies. Interestingly when items are connected in the brain, when there is a neuronal association between stored concepts, thinking of one of them will automatically trigger thinking of the associated concept according to the well-known neuroscience adage:

Neurons that wire together, fire together.

Could it be that our brains function much more like a Latent Semantic Analysis based program? That the logical inferences are only made after an association is detected, as a kind of proof-reading mechanism, which verifies whether the correlation is useful and in what way?

I already mentioned that we have no certainty that cause and effect, as we believe exist, really do exist. Quantum mechanics seems to reveal that the arrow of time can work in two directions and not necessarily only in the one direction we observe, as has become evident from Wheeler's delayed choice variant of the double slit experiment. This poses questions about the possibility of retro-causation: present events being caused by events in the future.

But there is also another explanation, which puts causality at a deeper level of reality. Modern physics is more and more venturing in the field of digital physics led by the Dutch physicist Verlinde,[9] which considers the universe as a kind of quantum computational substrate. In this model everything is information.

Gravity and entropy are not the direct consequences of the movements of corpuscular bodies but rather the consequence of information processing laws, algorithms at the deeper level of the computational substrate. Entropic gravitation results in proximity co-occurrence of corpuscular bodies such as planets, which maximises the ability to dissipate energy and to maximise entropy.

If this is true, then perhaps there is no direct causation at the macroscopic level we see, but the correlations we observe as causally linked are the consequence of a causation by algorithms functioning in the quantum computational substrate of the quantum vacuum. In fact this would imply that the whole universe is some kind of computer which uses a principle similar to the tendency to achieve proximity co-occurrence as in Latent Semantic Analysis. The universe could then perhaps also be some kind of mind. I am aware that this is slippery ice and pure speculation and full of the very logical fallacies I warned for, but I merely ask you to consider this possibility as an alternative explanation to causation.

The Latin word for reason is "ratio" which is probably not by coincidence linked to the English word ratio, which refers to quantitative relation between two amounts. The Greek word "logos", from which the word logic is derived, also means "reckoning". These etymological sources also point to a link between reasoning and (numerical or informational) reckoning.

Nevertheless, for the purpose of reasoning and dealing with the world around us, assuming the principle of causation at the macroscopic level is vital. We can only bring complexity about if our actions follow predictable patterns. So let's keep the metaphysical notions about causation only in back of our minds and, for practical reasons, put causal inference to our advantage.

The title of this book is *Is Intelligence an Algorithm?* If reasoning is part of intelligence, it must also be part of this algorithm. But I have shown that reasoning in AI seems to be unsuitable for context independent approaches. How can our natural intelligence then use reasoning as an algorithmic process which is context independent?

To understand this we must abstract the common features of the different modes of logic. What they have in common is that they all **compare specific instances with generalised rules**. So the intelligence algorithm will upon encountering a specific instance of an item seek in its database whether there is an ontological class it belongs to. It will compare specific with general using the rules of one of the modes of deduction, induction etc. and if a fit is achieved, recognise the item. As the item resonates with the general ontological structure of links and categories of the class, an association will be formed, not only metaphorically in the mind, but also literally at the level of the neurons, which will wire together and hence fire together.

If the item is new intelligence will look for similar items and build an ontology on the basis thereof and on the basis of the relations the new item has with existing ontologies: Thus it will cognise.

This part of intelligence is essentially a **comparison engine**, which can discriminate between items, weigh and classify them and draw conclusions if certain conditions are fulfilled (reminiscent of the often used "if... then... else" statements in computer algorithms). I'm telling you nothing new; in ancient India these aspects have already been described in great detail, for instance in the "Yoga sutras" by Patanjali,[10] which is an excellent treaty not only on the mental aspects of yoga but also on the workings of the mind. The intellect was called "Buddhi", the ability to discriminate between items was called "Viveka" and inference was called "Anumana".

The rules of logic when comparing specific with general are typically used to predict the outcome of future events and are thus mostly essential for planning, heuristics and problem solving. The specific strategies thereof will be the topic of a further chapter in this book.

Reasoning is also the most important part of "**Rhetoric**", the art of discourse in which you try to convince, to persuade your audience of your point of view. In a rhetorical argument, you can often start by giving **specific examples** of a **general principle** you wish to illustrate, or conversely you start by making a gener-alised assertion, which you then give a foundation by exempli-fying it with specific instances. It's clear that here you are using logic, most often induction and deduction. You are grounding an observation, just like Ben Goertzel[2] tries to do with his OpenCog and Novamente projects for the development of AGI (Artificial General Intelligence).

But there is more to rhetoric than logic alone. Rhetoric also appeals to psychological aspects; it appeals to your beliefs and morality or to your emotional propensity. Here we enter a more difficult area, which I will treat in more detail in a further chapter in this book. This area is more difficult because it diverges from traditional intelligence which is based on pattern recognition and logic. This area touches upon intuition, which I will try to illustrate is possibly a kind of hidden heuristic (a practical approach to problem solving without a guarantee for success; e.g. an educated guess).

In rhetoric based on emotions, the persuading orator can make an appeal to fear, which may block your more objective logical way of making sense.

The orator will try to seduce you to fall in the trap of logical fallacies, and come with evidence and facts, which may be true in a given situation, but not in all situations. On the basis of cherry-picking from probabilistically speaking insufficient information, he will try to make you apply your logic. Taken off guard by an emotional distraction, you may not apply your usual standard. And you may do the same when you try to convince someone else of your point of view. Perhaps with this chapter I am doing this with you. But at least by unveiling my mask I now give you the opportunity to seek truth for yourself.

This brings us to the issue of "truth". Truth as we experience it is a relative concept. For each event, different beholders have a different narrative which is often blurred by interpretations and coloured by beliefs and emotions. The same event can be told from very different perspectives, which at first glance appear contradictory and even mutually exclusive, but which in the end can be transcended from a higher perspective as relating to different parts of the same entity or process.

This is perhaps best illustrated by the Indian parable of the elephant: Several blind men touch an elephant to learn what it is. One touches the tail and concludes that it's a broom, another one touches a leg and concludes it's a pillar, a third touches the tusk and concludes it's a horn etc. Whereas from their own perspective none of them is really wrong, from the higher all-inclusive perspective they are all wrong to a certain extent and right to another extent. The dichotomies are resolved by a higher dimensional entity and perspective, which is the elephant, which transcends but does not exclude the partial perspectives.

Hence the famous quote by Nagarjuna:[11]

Anything is either true,
Or not true,
Or both true and not true,
Or neither true nor not true;
This is the Buddha's teaching.

As RA Wilson stated:[12] "What the thinker thinks, the prover will prove." In other words, you will always find proof and evidence to support your beliefs. Which means that if you really go to the bottom of this rabbit-hole, you cannot believe anything, because nothing is really certain. Hence Terence McKenna's famous quote:

Belief is a toxic and dangerous attitude toward reality. After all, if it's there it doesn't require your belief – and if it's not there why should you believe in it?

In addition we must realise that what we believe to be "reality" is a "virtual representation" of reality cooked up by our brains. Since our brains filter out a massive amount of information and since different people have different filtering capacities in their brains, how can we conclude that there is a common truth? There may be a kind of "consensus reality" as Bernardo Kastrup[13] calls it, which certain people agree upon, because their observations correspond. But the results of quantum mechanics have made clear that ultimately there is no "objective reality" out there. Your observation already changes the nature of existence, which is summarised in the famous adage: "When you change the way you look at things, the things you look at change."

So when it comes to truth, there are subjective truths and at best a consensus truth shared by a group of people. Plus the fact that we have all kinds of observational biases due to our personal and emotional background, we have cultural and linguistic biases and certain words are ambiguous homonyms. So no wonder we often fall prey to misunderstanding each other.

We already saw that logic cannot give us a solid foundation for our beliefs. Which does not mean we should discard reasoning: It is usually the only way we have to make sense of the world around us. But we must be vigilant to jump too quickly to conclusions, or to cast away someone else's perspective; we probably haven't seen the whole picture. So we must adopt a cautious pragmatic approach and replace our beliefs with probabilities and likelihoods. The more your intelligence increases, the less convinced you are of a certain specific point of view. Instead you will try to acquire the bird's-eye view, which puts different perspectives in context. You will try to find a meta-perspective.

So if someone is really certain about his or her case, beware! You may not be talking to a very intelligent person.

Chapter 4

Problem Solving

We have now arrived at the most important part of this book on Intelligence as an Algorithm: Problem Solving.

Thus far I have exposed my hypothesis that intelligence functions as a kind of algorithm in chapter 1 of this book. Chapter 2 related to cognition, (pattern) recognition and understanding. Chapter 3 explored the process of reasoning, which is necessary to come to identifications and conclusions and is also a tool in the problem-solving toolkit.

In this chapter I will discuss how we identify and formulate a problem, how we plan a so-called heuristic to solve it, how we carry out the solution and check if it fulfils our requirements.

Problems arise due to a discrepancy between the status quo of a system or living being, which has a deficiency of a certain kind and a desired future state without that deficiency. This discrepancy can be internal or external.

For instance, if a part is broken in a system (e.g. a motor) or if a living being is diseased, the system doesn't function as it is supposed to, it is functioning in a suboptimal manner, so there is an internal functional deficiency, which most often is caused by a defect at the structural level: a cogwheel may be broken, a fuse burnt or a gene may be deregulated to name a few.

External problems arise due to stimuli (or the absence of stimuli where there should be) from the environment and the relations of the system or living being thereto. A road may be bad hampering the movement of a vehicle, there may be a lack of resources in the environment for the system to be able to function or there may be a hostile opponent endangering survival, to name a few. The problem may also be of a psychological nature if we see that someone else is more successful than we are and we start to envy what the other has achieved. In that case too there is a deficiency of the status quo compared to the desired future state.

If we notice a system has a deficiency, we have to identify what exactly the deficiency is. This may be more difficult than these words suggest: The presence of a problem is often only apparent from the symptoms of a dysfunctional behaviour of the system. But the symptoms do not always reveal the structural cause of the problem.

So the first part of problem solving relates to the **identification of the problem**. From the symptoms we have to analyse the underlying causes to come to a correct diagnosis of the problem. We have to search for and explore which phenomena can cause the observed deficiency and check if indeed something is wrong with one or more of these phenomena. If so, that may be the cause of our problem and we have identified our problem. Correct diagnosis can be a daunting task, which requires a strategy to solve. In fact diagnosis of a problem is a problem of itself and requires the same strategical considerations as the stages of the problem-solving process which follow after the diagnosis.

To assess the problem we have to map the problem in terms of everything we know about it regarding symptoms, potential causes and a good description of the goal we wish to achieve. In fact we have to build an ontology (a descriptive list of features and relations describing an item) of the different aspects of the problem, which can be considered as the internal part of the problem description.

Once we have correctly assessed the problem we can start to devise a strategy to solve the problem. Fortunately, I don't have to develop the toolkit for solving problems from scratch. I can stand on the shoulders of a giant who has thoroughly mapped and described the process of problem solving in the framework of mathematical problems: George Pólya. But I realised these principles are applicable to any problem and not limited to mathematics only.

In his book *How to Solve It* published as early as 1945, Pólya[14] proposes a universal system of thinking, which can help you to solve any problem. Pólya describes this as a four-step process, the instructions of which can be called an algorithm in line with my thesis that intelligence functions as an algorithm.

The four steps are:

1) Understand the problem; 2) make a plan; 3) implement the plan; and 4) verify the result and see if it can be improved. I will discuss these enriched with ideas from my own insights.

As to understanding observations I have written in the second chapter about analysis, consideration and building an ontology of the observation. These principles also apply to the understanding of a problem.

In the analysis I propose that we must identify the 6 Ws (Who, What, Why, Where, When and hoW) first, in which the identification of the goal of the problem to be solved is the most important aspect. A good trick to see if you have understood the problem is to restate the problem in your own words. Ideally, you can abstract the essence of the problem in a simplified pictorial representation: an image or glyph.

You will not only profit from ontologising the problem in the form of a list of features, relations, laws, equations, conditions and restrictions, but you will improve your understanding even more if you can visualise the ontology in a diagram, a map.

The best way forward to map the ontology of the problem is to put the elements of greater relevance more central in the image in bigger characters and/or in a bigger frame. This generates a cloud of terminologies. Now you connect the different items with lines that represent their relations. The thickness of the lines can be indicative of the relative importance of the relation. This can result in a mapped ontology which can look like this:

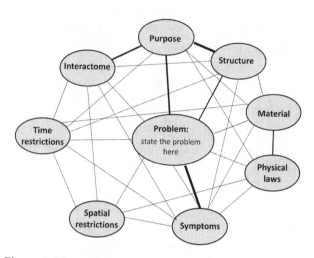

Figure 3: Mapped Ontology with weighted relation edges

Other types of graphical representations or schemes can also be useful, like a dendrogram, a grid etc.

If you want the distances between the different items to represent the relative importance (less important elements further away from the centre and the closer two items are to each other the more they are relevant to each other) this can be a difficult task. In the appendix to this chapter I will provide more details how you can do this. Ideally your ontology map shows all 6 Ws.

This ontology map of the problem can enable you to see whether you have enough information to solve the problem and if not prompt you to data mine for further essential information.

Vital to understanding a problem is that you understand all of the concepts and terminologies involved and, if you don't, that you update your information set with the missing meanings of these. You can also make a list of the questions which are still unanswered. Once you have understood the problem in line with Buckminster Fuller's[7] recipe, you can now articulate your understanding i.e. formulate the problem in a manner as detailed and precise as possible.

To Devise a Plan and Heuristics

We are now close to searching for a solution to the problem. It is important that we list the differences between the status quo and the desired solved state, ideally in terms of structures and functions and their associated effects. We can then start to devise the strategy to find a solution.

If it is a known problem we can adopt a conservative approach and simply consult books or databases to explain to us how we should solve it. However, most problems are unknown problems to us and we need a practical method to advance stepwise towards the solution.

This is the topic of "**Heuristics**" (Greek for "to find", "to discover"). A heuristic is a practical method to solve a problem, which method is not guaranteed to be perfect or optimal, but good enough to get us started, simplify the overload of data we have to deal with and allow us to progress.

You could call a heuristics a way of making an "**educated guess**". A heuristic is a way of exploring a potential solution space. Computer algorithms often use structured heuristics to solve a problem. But we do too, even if we may not always be aware of it. For instance, in board games such as "Battleship", you can use a strategy to launch your missiles in a more or less homogeneous distributed way over the 2D grid to explore the enemy's space. In the board game "Go" it is more advisable to first conquer the angles and edges of the board, whereas in "Chess" it is a general strategy that you should first strengthen your position in the centre. All these strategies of topological sequential advancement are types of heuristics.

Your intelligence and problem-solving skills may significantly improve once you realise that you have to make a plan with different stages to solve your problem and that choosing the best heuristic will warrant you the highest chances of success in the shortest time possible.

First we should try to find an analogous problem in a related technical field and see what type of solution was used to solve it. Here we can use the ontology of the problem and apply a search involving pattern recognition to find a problem which is identical or most similar to our problem.

Alternatively, we can try to see if we can find a similar result or effect in a neighbouring technical field even if the problem we try to solve was not explicitly mentioned there and see which structural features or configurations thereof yield the desired result or effect. We can then try to apply such a solution to solve our problem.

We can also try to find either a more general or a more specific known problem in the prior art.

If possible we should split our problem in smaller sub-problems which we can solve independently and later on integrate in an overall solution.

We can again "ontologise" and map the problem vis-à-vis such alternative solutions, giving a potential solution-scape (in imitation of the word landscape).

Some tools/considerations in Pólya's[14] toolkit regarding heuristics are the following:

Guess and check (this is a very simple random trial and error heuristic).

Eliminating possibilities which are likely to fail (for this you need to use reasoning as explained in chapter 3 of this book).

Use symmetries and complementarities. Pólya instructs us to look for patterns, draw pictures or make models.

Most importantly Pólya suggests considering special cases of the problem: It is easier to solve the problem for a concrete instance thereof and then to abstract generalised rules to solve the general problem you wish to solve rather than starting to solve the general problem from scratch. Pólya calls this "solve a simpler problem first".

Another heuristic he proposes is to "work your way back" from the desired solved state. A specific example thereof is "reverse engineering". Reverse engineering is the situation where you find something which solves your problem but in which case it is like a black box to you: You don't know how it functions and you cannot make it yourself. For instance you find an alien spaceship with great technology and you try to figure out how it is made and how it functions by disassembling it, by dissecting and analysing it, to give you clues how to put it back together to make it work. Biotechnology often works that way; we have to dissect the complex systems in their simpler parts and figure out how they synergistically function together. In such cases "reverse engineering" is helpful.

However, often we don't have the luxury of finding such a working solution and we only have a mental conceptualisation of the desired result. Even then you can try to work your way back. In chemistry scientists use the technique of "retrosynthesis", which is the best heuristic to figure out how to make a complex desired molecule by figuring out which simpler molecule with only one or two differences could precede the desired molecule and do this repeatedly at each stage for each identified simpler molecule until you arrive at two known starting materials, which you can purchase.

To solve a problem we can essentially only perform three types of actions: We can add something, we can take something away or we can modify something. A modification or substitution of an item for another basically involves taking away the item and replacing it with a similar item. Advancing in a territory means taking away your pawns from one location and adding them to a different position. Substitution essentially is successive subtraction and addition.

Certain problems can only be solved by omitting a constituent, which results in a simplification which yields better results. Other problems require the addition or substitution of an element. It is important to know that such additions or substitutions can have an unforeseen synergistic effect, which is called strong emergence.

Many inventions and scientific advances involve the observation of such a surprising effect. In fact often scientists make an observation they were not looking for at all, but which is even more interesting than what they were looking for. This is called "Serendipity". The resulting scientific publication is nevertheless often presented as if this was exactly what they were looking for. Don't let yourself be fooled: They often reverse engineer the hypothesis and scientific process by which they should have arrived at the serendipitous result as if it was a planned problem-solving process relating to an existing hypothesis or problem.

The exploration of a solution space of potential solutions, of alternatives to the status quo, is a "Screening" process. The way you strategically plan your screening protocol by prioritising certain alternative types of solutions over others is your heuristic. It often involves eliminating directions which are likely to be unsuccessful and this elimination process is called "Pruning". Since prioritisation is a daunting task itself I will give you a strategy for setting priorities in the appendix to this chapter.

To follow a conservative known heuristic can be the start of a problem-solving strategy. However, if this fails more creative strategies may be needed. An out of the box thinking may be needed where the exploring problem solver ventures into unknown territories, like looking for a similar problem or effect or ontology in unrelated technical fields. Or to see if a different ontology has nonetheless the same map structure, which is indicative of a similar functional behaviour despite the content differences.

Later in this chapter, I will show in what way evolutionary living systems explore creative solutions.

Once a promising potential solution has been identified, the remainder of the problem-solving process, the actual implementation, must be planned timewise and subsequently be carried out. It is important to set **milestones** of what should be achieved at a given stage in the process of implementing the solution. This allows for monitoring whether one is progressing in the right direction towards the solution and allows for adjusting the process, steering away or even returning to an earlier stage of the development if it is drifting away from the desired solution. In other words we need an operational **feedback check mechanism** involving intermediate measurements and checks for fulfilment of conditions, restrictions, equations and laws as intermediate results.

Before implementing the solution in a real situation it can be advisable to first simulate it in a computer if possible and if a mathematical model is available or can be made.

There are quite a few mathematical computer heuristics, like the Random Walk Monte Carlo method. To treat these heuristics falls outside the scope of this chapter and is of no interest to the educated scholar who has a better understanding thereof than I do. This book is for the layman using every day calculation tools such as Excel. There are two numerical methods in Excel, which many people don't know solve complex problems, that I would like to share with you: "Goal seek" and "Solver". They are very easy to use and useful in recalculating mortgage changes. Numerical methods such as Goal seek use successive increments of value. Solver is a more versatile version of Goal seek, which can be used to optimise when facing multiple constraints.

Once a solution has been reached another check is carried out to see if the solution is satisfactory under all thinkable conditions and to see if a better more optimal solution can be envisaged. You can also store this problem and solution in a problem-solving database, to accelerate future problem solving.

The different possibilities I have suggested for problem solving can be implemented in an organised and prioritised way by the intelligence algorithm: Conservative heuristics first followed by evermore daring creative ones if success does not ensue.

Now I wish to return to the exploration of creative solutions in biological societies such as beehives, anthills or bacterial colonies. In these societies there is a group of mostly workers, which implement a conservative strategy to maintain the status quo. These are called the "conformity enforcers". The system has also "inner judges" that monitor whether the group standard or morality is maintained. However, in times of difficulties such as a lack of resources individualistic explorers are needed, who venture in unknown territories and seek for alternative resources: these are the diversity generators. If they do find greener pastures and hit the proverbial jackpot, their success is celebrated beyond measure. The system can start to boom again and the group of resource shifters will exploit the new resources or educate the conformity enforcers to do so. Exhaustion of resources gives rise to "depressions" in the system, resulting in putting the system on a low metabolism regime or even shutting down the system entirely in a hibernation mode, safeguarding the information in well protected spores, which can bloom again in the future under more fruitful conditions.

Biological and physical systems are thus part of what Howard Bloom[5] calls the "evolutionary search engine". This evolutionary search engine is a natural intelligence that solves problems by a screening and pruning process and thus implements a solution. Intelligent systems mimic each other, absorb each other, niche or arrive at a symbiosis. They can also eliminate contenders in an intergroup tournament or mutate into something different. These are so-called "fission-fusion" strategies. The result of such an intergroup tournament need not be dominance or extermination. Sometimes they arrive at a kind of exchange of features, a biological commerce as a form of symbiosis, which shows that even natural intelligence is capable of finding the so-called "Nash equilibrium".

The mathematician John Nash[15] realised that the overall result of a cooperation between parties can be better or have a higher probability of success than a competition between the parties. This resulted in his theory about bargaining as a part of "game theory" and in the so-called "Nash equilibrium", which is the equilibrium reflecting the best overall result. He realised this when he and his male fellow students met a group of female students. If all the boys would have competed for the most beautiful girl, probably no one would end up being successful (in picking up a girl) in the group or at best one of them. If instead they agreed to ignore the most beautiful girl in the group and not to poach each other's territory, chances were that more of them would be successful to pick up a girl.

Cooperation can not only lead to bargains with an overall better outcome for the participants together and avoid that someone is left out; it can also yield synergistic effects. It is only not worthwhile if the cooperation would slow down the process, which can happen if you have too many participants for the work to be done, which is known as the law of diminishing returns.

With regard to human intelligence, as a part of our mental heuristics we run a plethora of virtual scenarios in our brains to solve problems. A bottleneck to be avoided in this process is to start ruminating, getting self-absorbed and running the same scenarios over and over again. To avoid this it is extremely useful to write down the different scenarios you thought of, ontologise and analyse them and weigh their respective relevancies so as to come to a conclusion.

Intelligence in human social, psychological and emotional settings, however, involves a different type of intelligence, an additional degree of complexity and hidden motivations I haven't touched upon in this chapter. These apparently more irrational considerations at stake in humans will be the topic of a later chapter in this book. Noteworthy, when we solve problems we often use an elusive heuristic called intuition, which cannot be pinned down to a specific sequence of preprogrammed steps (and can hence not be considered to be "algorithmic"). This will be the topic of the last chapter of this book.

Appendix:

Prioritisation

The easiest way to prioritise a number of actions to be taken is to assess their relative relevance of priorities in pairs in a grid. If the pairwise priorities are like this (the symbol > here means "having priority over") A>D, B>A, B>C, B>D, C>A, C>D, then you can represent this in the following grid:

	A	B	C	D	Sum
A	0	0	0	1	1
B	1	0	1	1	3
C	1	0	0	1	2
D	0	0	0	0	0

Equal relevance scores a 0. If a horizontal action is more important than a vertical one (e.g. B has priority over A), a 1 is assigned. If a vertical action is more important than a horizontal one a 0 is assigned (e.g. C has priority over A). The higher the horizontal sum of the values in a row, the higher priority of the item. Here the result is B>C>A>D. This we could have seen immediately without a grid for four items, but if you have 15 items to put in order, this is a very fast way to prioritise.

Topological relevance mapping

Now in addition to the priority of the items we need values of the importance of the pairwise relations e.g. AB=2, AC=3, AD=5, BC=4, BD=5, CD=3. In this example the priority order is given to be A>B>C>D. How to put this in a 2D map, where high priority items are closer to the origin and the lower the priority the further away the item is? We first draw a series of concentric circles with equal distances of length 1. We place our most important item A at the centre. Then we take a horizontal rod of length 4, which is the length of BC and shift it over the screen until its respective ends touch the second and the third circle simultaneously (which represent AB=2 and AC=3 respectively). Then with rods of length 3 and 5, corresponding to CD respectively, AC or BC, we position D at the right position to give the following result:

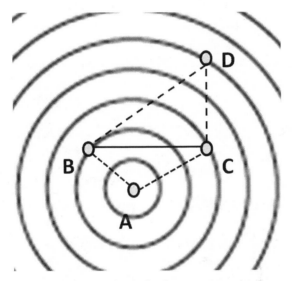

Figure 4: Topological relevance mapping

Thus we have been able to map items according to their priorities and with pairwise distances representing their relative relevance.

Chapter 5

A Template for Writing and Organising Thought

This chapter presents you a structured template for writing informative articles. It is also a tool to organise your thoughts on a certain topic. The clarity this will bring has also potential to increase your intelligence i.e. your ability to solve complex problems in complex environments. Essentially, it provides a structured means to create an ontology about a topic, to analyse a problem and to provide a solution thereto.

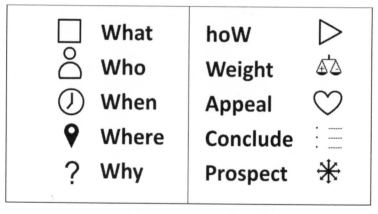

Figure 5: 7WACP scheme for drafting articles

It is especially for those authors who have a bit of a difficulty in presenting a topic in a logical structured way that I have designed the 7WACP template. It will provide you with a guideline and following order how to progress and how to conclude your story. I have also gathered a set of useful icons or glyphs (see image above) that will help you to remember and apply this structure systematically, if you wish. Note that this template is only a suggestion, you can write in whatever way you like, but if you need some structure, it can be quite helpful. You can also use only certain elements of it.

When you use the 7WACP template, you structure your argument by answering seven "W" questions as regards What, Who, When, Where, Why, hoW and Weight, which you need to answer to come to an "Appeal" (A), a "Conclusion" (C) and future "Prospects" (P) (see figure 5).

What

In the "What" section you summarise the topic you are going to discuss. You tell your audience in a few words what you are going to tell them. If the topic is ambiguous, you can include a disclaimer indicating what this topic is not about. For instance in this chapter I only provide you with a template for the content and structure of informative articles, but not for poetry or story-telling. This section can include an explanation of a hypothesis you try to prove.

Who

In the "Who" section you identify your audience. For instance the present chapter is written for those newbies who have a bit of difficulty in organising their thoughts in a structured way for an article. Here you try to seduce your readers to continue to read by promising them they will get a conclusion and answer to the questions they are looking for to be answered.

Background: When, Where and Open-scape

In the background section you give a historical context of your topic, like similar topics that were previously known. If possible you provide a geographical setting or a geometrical configuration in which your topic is positioned. This is your background art, for which you'll have to do some research and/or some data mining. You can also indicate the relations of the items in your topic with known items it interacts with. This is the so-called "open-scape" or "interactome" of your topic. You can summarise drawbacks in the items of the historical context, which are overcome by the topic you present.

Why

You can now indicate in the "Why" section what brought you to write this topic and why it solves the problems remaining in the background art. You thus indicate the purpose of your essay. For instance others have suggested how to improve your writing by layout, formatting and editing skills, but no one has come up with a structured template for blogging. This chapter aims to provide you with that structure.

HoW

In the "How" section you give the actual body of your argument. You can start by listing the structural features (form, material, and design), functions and purpose related to the items of your topic. You can also describe the relations of the items with open-scape items. Certain topics are more abstract or relate to human interactions, but even the features of these can be listed. This provides a framework for understanding your topic. This is the provision of a so-called "ontology". For this you may again need to do some research and/or data mining.

From this you can provide a further list of the differences between your topic/argument and what was previously known about it.

You can now list various known ways to bring about these differences and indicate which way you have chosen and why.

Weight and Integrate

As an intrinsic part of this "how-method", you will now have to "Weight" the pros and cons of those differences. Ideally you can list what effects are associated with those differences. You can repeat what problem was at stake and detail this more in the light of these differences and explain how these differences overcome this problem. For this you need a line of **reasoning**, in which you illustrate by giving one or more **specific examples** of how your hypothesis is proven or how your problem is solved, and from these you abstract generalised patterns. You have now **integrated** all the information to yield the conclusion. Throughout the body of your article (i.e. in every section) it can be useful to illustrate **abstract general ideas** with **concrete** specific examples.

You can also list alternatives to your hypothesis or solution and again via weighting pros and cons argue why these alternatives are less likely or less useful.

Appeal

You may try to win your audience by making an "Appeal" to them. Why this was important for them or why they should support your argument. Here you can use some psychology and put yourself in a specific role. The role can be an educator, a parent, a peer, a manager, an employee, a child, someone seeking support. You can also make an appeal to the valuation of aesthetics, originality and creativity of your hypothesis and/or solution (either your hypothesis can be original and the solution known from different areas or your hypothesis can be known, but your solution can still be original. There are also original hypotheses with original solutions/argumentations).

- ------
- ------
- ------

Conclude

Now you can conclude by summarising what your hypothesis or problem was; which solution you have chosen and why this is appealing to certain people. Basically, you tell your audience what you have told them.

Prospects

Finally you end with a list of possible future prospects, suggestions and developments which can arise as a result of your ideas. You can also list potential alternative applications of your concept.

You have now seen how a set of 7 simple W questions combined with an appeal (A), conclusion (C) and prospects (P) give an easy-to-implement format for blogging informative articles.

If needed you end by thanking your audience for reading your article and invite them for comments and questions.

The scheme I have just proposed can also be used for ontology building, problem solving, managing and giving presentations. And if you are lucky enough, you increase your general awareness and intelligence of the world around you.

Chapter 6

The intelligence of Emotions

In this chapter I will discuss how emotions are part of the natural intelligence algorithm. I will also indicate how an artificial equivalent of emotions could be of benefit to an artificial intelligence. The topic of how to deal in an intelligent way with your emotions will be the subject of the next chapter.

Background

In the previous chapters I have exposed my hypothesis that intelligence functions as a kind of algorithm (see chapter 1). Chapter 2 related to cognition, (pattern) recognition and understanding. Chapter 3 explored the process of reasoning, which is necessary to come to identifications and conclusions and is also a tool in the problem-solving toolkit. In chapter 4, I discussed how we identify and formulate a problem; how we plan a so-called heuristic to solve it, how we carry out the solution and check if it fulfils our requirements.

This chapter is written for those people who wish to understand more about the innate intelligence associated with emotions, ultimately in order to be able to deal better with emotions and not be overwhelmed by them (next chapter).

Many psychologists have written about emotional intelligence or have described what emotions are. It is not my purpose to summarise what they have said. Rather I would like to show how emotions fit in the intelligence algorithm I have been describing before. Note that nothing of what I assert is necessarily so. I am formulating a hypothesis and will try to convince you of its plausibility.

In human interactions decisions sometimes seem to be based on irrational considerations, which appear to deviate from the rational process of the intelligence algorithm as set out in the previous chapters. In this chapter I will try to argue that emotions are an integral and vital part of the intelligence algorithm. What appears to be irrational is merely a different aspect of intelligence in operation. So let's try to decipher what kind of mechanism or algorithm is involved in emotions.

Where and when

Emotions are formed and stored in the part of the so-called "limbic system" in the brain called the "basal ganglia", more specifically in a part thereof called the "amygdala". It is also in the basal ganglia that fixed action patterns of immediate responses to emotions are stored and released whenever necessary.

When we make an observation the first percept is essentially neutral. It is only when we interpret the thought associated with the observation that we colour the observation, that we interpret it. This thought-based identification called apperception then leads to the triggering of emotions.

Imagine you see a wolf. First your brain identifies that it is a wolf. That part of the apperception is still neutral. It is only when we interpret the context that an emotion can be released. If it is a wolf in a book or in a cage, our first emotions triggered will not be fear. If it is a wolf in nature walking freely at a visible distance from us, our brain will interpret this specific combination of features as being related to imminent danger. This will go so fast that these thoughts won't even be articulated in words, but the non-verbal interpretation needs to happen before the basal ganglia can release a so-called fixed action pattern, which we call an emotion, in this example fear.

So the questions, as to why we have emotions and what the evolutionary advantage is, can be answered quite simply: because we can't afford to go through the rational steps of ontologising, weighing and finding an appropriate heuristic in the normal way. The emotion is like a flickering light on our dashboard, a monitor warning us that there is something wrong, that action is imminent and that we can't reflect on an action but need to follow a fixed action pattern.

However, sometimes emotions give false, deceptive signals and how to recognise these false signals will be the topic for the next chapter. The next chapter will also give tips on how to control emotions and use them to our advantage.

To understand emotions we have to try to build a more complete ontology of them and abstract generalised patterns therefrom. This is the topic of the present chapter.

What are emotions?

As I just identified, emotions are the fixed action patterns that are released as a result of internal and external **monitoring**: It is a dashboard **indicator** that lights up to make us aware of an imminent situation, which may have repercussions for the system we are.

Emotions show whether a desired state has been reached or not in comparison to the preceding (status quo) state. Emotions also show whether it is likely that a desired state will be reached or can be maintained, which we call anticipation. Let me illustrate this with regard to a few basic emotions: fear, sadness, anger, joy, excitement, jealousy and awe.

- Fear is the anticipation that a desired state may not be reached or cannot be maintained.
- Sadness is the finding that a desired state has not been and/or cannot be reached or maintained.
- Anger arises when the action of a person or thing forms an obstacle to reaching or maintaining a desired state.
- Joy is the finding that a desired state has been reached or maintained.
- Excitement is anticipation that a desired state is likely to be

reached.

- Jealousy arises when we compare our status quo with that of someone else and start to desire that state with the uncomfortable anticipation that we might be incapable of reaching that state. It is also an indicator that there may be a better state than the one we have reached.
- Awe arises when we admire the better state or achievement of someone else without desiring to reach that state, because we are convinced that we cannot reach that state anyway.

Thus here above we have classified emotions in terms of positive and negative (possible) outcomes.

Emotions can therefore be defined as indicators to notify us of our state in comparison to a different (possible) state. In this sense they are useful as elements for an algorithm. Algorithms also need to monitor and assess if a given state has been reached and inform the user and itself by feedback so as to be able to progress to the next step. In designing artificial intelligence we also need such indicators. We may not call these emotions, but in fact these internal checks are artificial emotions.

Emotions are therefore useful elements which can help a system to find its way in a complex environment to achieve a complex goal, which is exactly Ben Goertzel's[2] definition of intelligence. Emotions protect us; show us when to advance and when to retreat.

Personality type ontology

Most people are capable of having the same types of emotions, but not necessarily in the same degree or under the same circumstances. In most cases men and women particularly differ in this respect. Then there are stratifications of people in different types such as being based on their level of accomplishment (in analogy to Maslow's pyramid), RA Wilson's[12] typing of viscerotonic, musculotonic and cerebrotonic types and the DiSC profile by Marston.[16]

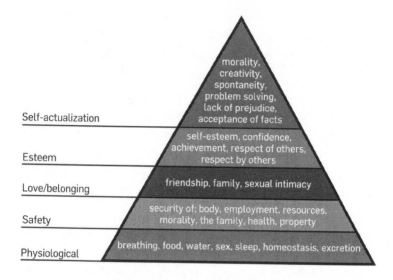

Figure 6: Maslow's pyramid[17] of needs. By Saul McLeod – http://www.simplypsychology.org/maslow.html, CC BY-SA 4.0, https://commons.wikimedia.org/w/index.php?curid=47784797.

Since emotions are linked to needs and desires a correspondence map with hierarchy of Maslow's pyramid[17] can be made. Maslow's pyramid describes needs and desires in 5 levels:

- **Physiological** needs such as breathing, food, water, shelter, clothing, sleep.
- **Safety and Security** needs such as health, employment, property, family and social stability.
- **Love and Belonging** needs such as friendship, family, intimacy and a sense of connection.
- **Self-esteem** needs such as confidence, achievement, respect of others, being a unique individual.
- **Self-actualisation** needs such as morality, creativity, meaning and inner potential.

Fear relates to the level of physical needs such as shelter, food etc. At the next level of economic and social safety anger is the more prevailing emotion. The next level of love and belonging corresponds to joy. The level of esteem corresponds to excitement and the level of self-actualisation to serenity and equanimity. Feelings of superiority and inferiority, dominance and submission are associated with the levels of safety and esteem.

Another useful scheme for stereotyping people is the DiSC[16] scheme used to type managers. Real people, however, often are a mix of the stereotypes I am going to describe, so don't take these stereotypes too seriously. Don't think I judge people in this way either, these are just traditional stereotypes described in various sources. It is by no means my intent to be condescending toward anyone.

The DiSC model stratifies people along two axes: The introvert-extrovert axis and the people-versus task-oriented personalities. DiSC stands for Dominance, Influence, Steadiness and Conscientiousness.

Figure 7: DiSC model. Available online from: http://eqevents

.com.au/how-to-win-friends-and-influence-people/. Reprinted with permission from C. Pollard from eQ events.

Dominant personalities are extroverted and task oriented. A typical military dictator is a good example of this type. They correspond to Wilson's "musculotonic" type and the noblemen in the ancient caste system. Their behaviour is intimidating, based on feelings of power and superiority. They may use force to achieve their goals not always taking into account the effects on other people.

Influence type personalities are also extroverted but are more people oriented. They are said to have better communication skills and inspire people. They often strive for progress. They correspond to the class of businessmen, the city dwellers or bourgeoisie in the caste system. They are excited, full of energy seeking enjoyment in life. Their physical constitution in Wilson's terms would be quite normal. Their emotional approach is warm, welcoming and not hostile.

Steadiness represents introverted friendly naive people. Without wishing to denigrate there is said to be a certain correspondence to the working class including farmers here. They are said to be fearful and suspicious and rather would like to keep things as they are. Some of them suffer from an inferiority complex. Their body constitution can be "viscerotonic" in Wilson's terms (i.e. a bit stout). They are more people than task oriented. In the ancient caste system they corresponded to the servants.

Compliance type conscientious personalities are the academic people, the clergy, scientists and geeks. Their body type is "cerebrotonic" in Wilson's[12] terms i.e. thin, long body with a high head. Typically introverted and task oriented, they like to do things alone. They don't understand and show emotions easily.

They can exhibit a bit autistic behaviour and show frustration if they can't get their message across. They will only show excitement when their mental creations are successful. They are often perplexed about the behaviour of other types, which they consider irrational, and are suspicious and reserved towards emotive personalities.

As said before real people are a mix of these stereotypes and depending on their sociocultural position in the pecking order they may have a certain proclivity towards certain attitudes and emotions that have been imposed on them by their cultural heritage and education. Each of these groups has figured out a survival strategy belonging to their niche in the social patchwork. Their specific emotive subcategory patterns warrant them a certain degree of success in their niche.

Natural intelligence has selected certain fixed emotive release action patterns, which they believe will yield them the desired result. Their different desires as regards their degree of extroversion and people vs. task orientation result in different sometimes colliding strategies. When people of different groups/classes meet in a complex context many misunderstandings can arise due to different sociocultural phenotypical programming.

Each group has different types of filters to classify and interpret observations. By the intergroup tournament, i.e. the competition and interaction between the groups, a natural pecking order can arise.

How emotions manifest

Emotions often manifest physically in the form of increased muscular tension, pain in the belly, crying, numbing, wetting one's pants etc.

Why we have emotions

The reason why we have emotions as already said is to warn us, make us aware that we have to take action. Without emotions there is no monitoring of internal states. Without such monitoring imminent danger would not be recognised. If there is no care for others – as expressed in the emotion of love – no societal complexity can arise. This is why mammals are more successful than reptiles. Mammals have a limbic system capable of emotions. Emotions are there because there is no time for a deep heuristic search to evaluate and consider a situation. The slower rational cortical process must sometimes be overwritten by fixed action patterns in the amygdala. Hormones also play a role in bypassing the cortical neuronal information flux.

Process ontology

I'd also like to ontologically map emotions with regard to an eight-phase cycle of perception, thoughts, action, and feedback

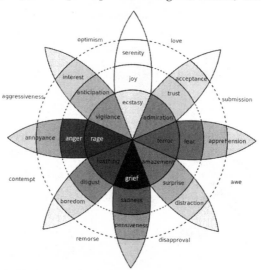

Figure 8: Plutchik's wheel of emotions. By Machine Elf 1735 – Own work, Public Domain: https://commons.wikimedia.org/w/index .php?curid=13285286.

corresponding to the intelligence algorithm described in the previous chapters. The emotional classification by Plutchik[18] can be quite useful for this purpose. Emotions mentioned in the analysis below derive from Plutchik's classification.

Phase 1: Perception / Scan
A phenomenon, a stimulus (e.g. object, sound, and light) enters the brain via the sensory organs. This I'll call a perception stimulus.
   Emotion: interest-vigilance.

Phase 2: Identification / Search
This phase involves first a descriptive, a feature analysis and an interpretation sub-phase. The brain (lower thought processes) will then start searching using association laws for: 1) similarities, analogies and resemblances to mental objects stored in the memory (compare); 2) opposites, contrasts, differences (distinguish); 3) spatial associations (locate); 4) temporal associations (time). Thus the brain will be able to organise, classify, categorise and ontologise the phenomenon.
   Emotion: anticipation-optimism.

Phase 3: Cognition / Conclusion / Judgement / Apperception (spirit)
On the basis of the most probable result of Phase 2 (less probable pathways unlikely to rapidly yield success will be discarded) a conclusion will be reached and the phenomenon will be recognised or at least, if it turns out to be new, it will be associated with the phenomenon it resembles most. The conclusion is followed by a judgement about the phenomenon which finalises the conclusive phase. With judgement here I do not mean as a first consideration defining whether something is morally good or bad, but rather the following: An emotion is observed, classified as pleasant or unpleasant, a threat is observed, a

problem is noticed or indifference is concluded. The judgement phase can also involve an abstraction phase (higher thought processes) wherein the essence is distilled. This essence can be an archetypical form, but also an underlying principle, an algorithm.

Emotions that follow this phase when positive vary from serenity to joy or even ecstasy; when negative they vary from fear to terror or from pensiveness, via sadness to grief; from annoyance to anger.

Phase 4: Planning phase: Problem definition / Design solution (solve)

The outcome of Phase 3 will lead to the need to take action if a problem must be solved or to non-action, if the result is indifference or negative. Even if there are not sufficient resources to solve the problem, the brain will necessarily go through these phases of problem/solution evaluation, even if the outcome is that the resources (energy and/or capabilities) are insufficient to provide the solution. The brain can only acquiesce to that outcome if it has been able to evaluate that the resources are insufficient.

The problem must first properly be defined and then a solution designing phase can take place, in which again a search analogous to Phase 2 using association laws is used: If a proper or seemingly analogous solution is already available, this will be used or at least taken as starting point. If no good results are available, then the evaluation may result in the assessment that the resources are insufficient (disappointment). Once a solution pathway has been chosen, the details of its future implementation will be planned.

This phase also goes through an evaluation of the degree of necessity/urgency and the availability of resources, the product of which must be bigger than a certain threshold in order for action to be taken. Furthermore, this phase also involves a moral assessment as to the desirability of attaining the solution-to-the-problem or fulfilment of the need/desire. Here the modal principles as described by Kant come into play.

Accompanying emotions: acceptance-trust (positive); disappointment – disgust-loathing-annoyance-anger (negative).

Phase 5: Implementation phase (score) / Behaviour
Firstly a decision will be taken as to whether the brain will use the willpower (Ego) to implement the envisaged solution. Note that the details of the solution may not be known to the brain; it will rely on intuition, a hunch, a "Fingerspitzengefühl" that potentially a right solution has been chosen or at least one that is worthwhile trying. Via the organs of action the solution will be implemented in the material world or at least its first stages. This is where the mind will make itself known to the material world and where an action materialises.

Accompanying emotion: submission (of the material world to its needs)-apprehension-fear (for failing).

Phase 6: Perceive reaction / Observe result / Effect
The feedback loop starts here. The mind will perceive the results of what has been materialised in action: the reaction of an interlocutor, a strike of a pencil etc. This will strike the brain positively or negatively.

Emotion: surprise-amazement-awe.

Phase 7: Evaluation / Sense
The result will now be evaluated, pondered and meditated upon. A search similar to the one in Phases 2 and 4 will classify the result and the result will then be estimated/measured by the discriminating faculty of the intellect. The outcome can be disappointment or satisfaction.

Emotion: negative: pensiveness-sadness-grief; positive: serenity-joy-ecstasy.

Phase 8: Re-evaluate / Reject or reinforce (serve)

On the basis of the result of judgement by the intellect, the emotions will react negatively by rejection of the action one undertook or positively by admiration of what has been achieved, leading to a reinforcement of the chosen solution. Negatively, the disgust and loathing of the failure may even turn into contempt of oneself and lead to annoyance, anger or rage. Annoyance will lead to abandonment of seeking a solution to the problem, anger or rage, may be an incentive to look for a totally different solution. Positively a virtuous cycle is entered and the solution may become a more permanent tool. Note that this eight-phase scheme encompasses the well-known 5 step scheme of: stimulus – cognition – feeling – behaviour – effect, but I do not see the emotions as separate from the cognition, rather they are part of it; and furthermore, I reckon that each phase has its inherent emotions.

Conclusion

We have seen how the natural intelligence algorithm uses emotions as indicators of (potential) achievement of desired states. How emotions depend on circumstances and sociocultural programming. How emotions result in necessary fixed action patterns, which may seem irrational but follow a prepro-grammed sequence directed at reacting to imminent situations. That sometimes emotions can become long-term problems, such as in long-term grief, points to the fact that emotions can derail beyond what they are naturally intended for. How to deal with the process of such undesired derailed emotions (deceptive messages) is *inter alia* the topic of the next chapter, which deals with how we should navigate the emotional quagmire of false signals.

Prospects

The ontology of emotions and the notion of emotions as

indicators of internal states should be further explored for the development of artificial intelligence. If we can create emotional machines capable of compassion, perhaps we need not be so afraid of the dystopian doom scenarios that are predicted upon the advent of strong AI. The machines may then not be inclined to wipe us out. Moreover I predict it is even vital to the design of artificial consciousness that there are internal monitoring feedback mechanisms. For that is what consciousness actually is: A feedback mechanism that integrates information to update the status of the entity in question.

Chapter 7

Emotional Intelligence

In this chapter I will discuss how to navigate the raging sea of our emotions. The chapter presents a collection of techniques to control our emotions deriving from psychology, management skills and practical teachings by spiritual masters stripped from their religious context.

Background

In the previous chapters in this book I have exposed my hypothesis that intelligence functions as a kind of algorithm (see chapter 1 of this book). Chapter 2 related to cognition, (pattern) recognition and understanding. Chapter 3 explored the process of reasoning, which is necessary to come to identifications and conclusions and is also a tool in the problem-solving toolkit. In chapter 4, I discussed how we identify and formulate a problem; how we plan a so-called heuristic to solve it, how we carry out the solution and check if it fulfils our requirements. Chapter 6 discussed how emotions are indicators which are part of the natural intelligence algorithm and which indicate whether a desired state has been achieved or maintained.

As promised I will now first embark on the topic of how to navigate in the quagmire and quicksand of emotions. I will not discuss the complete literature by psychologists on this topic, but rather cherry-pick some valuable insights from psychology, summarise patterns that I have distilled from my life experience and mention a few teachings from certain so-called "enlightened people" which I have collected as valuable tools.

Please note that if in this chapter I appear too pedagogic or even pedantic, this is clearly not my intention. You make ask yourself: "Who is this guy that he claims to know what intelligence is and to know how emotions can be mastered?" I don't. I interrogate. I speak from experience insofar as I have suffered my share in life and I have actively searched for solutions to reduce this suffering. This chapter is a summary of those techniques I have read about and practised and found to be effective in reducing my suffering. I don't claim to master these techniques perfectly, but whenever I do apply them, they do result in a certain relief. This does not mean that I perfectly master my emotions. I can sometimes react quite primary, but whenever I am aware enough to apply the principles below, they do solve my problems. This is not an exhaustive treaty either, but a useful rule-of-thumb collection.

As discussed in the previous chapter there is nothing wrong with emotions per se; they are useful indicators that update our knowledge of our status quo. However, we are not always interpreting the readout of these indicators correctly and sometimes we trigger a self-reinforcing hysterical or paranoid action from which we suffer. It is this overreacting, this mental exaggeration that causes us to suffer.

To navigate our emotions we must therefore learn how to adopt either a social skill strategy to adapt rapidly to give context or a more equanimous attitude, so that we can avoid being overwhelmed and dragged in to a vicious circle of feedforward emotional escalation.

I will start with the more specific toolbox of social skills and then discuss the more general toolbox for learning equanimity. Finally, I will dedicate a short paragraph to artificial equivalents thereof needed to avoid the creation of a "nutbot" as an introduction to the topic of the next chapter.

Social skill toolbox

The problem with many people is that they identify only with a limited group of other people. They often identify solely with people from the same sociocultural background. When they have to interact with people from different sociocultural niches, problems start to arise. Even if they officially speak the same language, they do not speak the same sociocultural language. Each group has its own filters for reality; sees reality from a limited perspective. Due to these differences misunderstandings arise, which give rise to unpleasant emotions such as fear, hatred, anger etc.

In order to be able to intelligently navigate in the sociocultural conundrum of the hivemind of humanity it is important that we learn to transcend the limitations of our safe sociocultural niche. It is important that we learn to adopt the bird's-eye view.

Mimicking

Ideally we can put ourselves above the different categories and castes we have seen in chapter 5 of this book, which gave a stratification of personality types. Ideally we become like chameleons that proverbially behave like a Greek with the Greek and like a Jew with the Jewish. Cunningly mimicking the behaviour of the person you are interacting with in order to gain his or her trust. Make them believe you are one of them; use their language and habits in order to ultimately be able to impose your goals once you're "in". Sounds pretty much like what politicians do, isn't it?

Complementing

Apart from mimicking we can also complement behaviour. In psychological models, often mention is made of Parent, Adult and Child roles that can be adopted. In more detail, a so-called caring Nurturing Parent naturally talks to a Natural Child, and a Controlling Parent to an Adaptive Child.

Even adults assume these roles in the form of a natural pecking order process. In fact these parts of our personality can be *evoked* by the opposite. When people are assuming to be talking to someone in a given role, but that person adopts a different role, problems and misunderstandings arise. A controlling parent wishes to see submission by an adaptive child. If your interlocutor assumes a different role, rebellion can arise. In the scheme below this is illustrated:

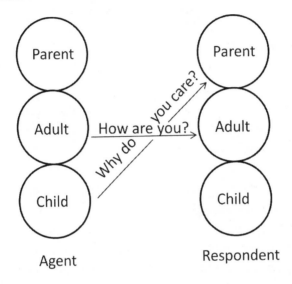

Figure 9: Eric Berne's[19] "Transactional Analysis".

When lines cross in this diagram, problems arise: In the example given in the diagram a person addresses a peer on the adult level and asks, "How are you?" The interlocutor, who is not feeling well, interprets this as a criticism from someone in the parent role and adopts a rebellious child role by cynically responding, "Why do you care?"

To be aware of such stratifications and roles can improve your ability to deal with interpersonal emotion exchange tremendously. Don't interpret the reply too quickly and don't respond in a primary manner, but analyse which level of interaction is at stake.

It is often useful to verify whether you have understood the intention correctly, by asking, "Did you mean…" This will allow you to avoid having difficult feelings about presumed intentions of others, which may not even be there.

If there is a problem, it is often one of feelings of inferiority. Even if someone tries to behave as your superior, this only betrays that he/she actually is insecure about his/her position in the pecking order, which itself already reveals a feeling of inferiority.

You are not inferior to anyone, nor are you superior to anyone. We all just have a different set of abilities.

Improve your Linguistic skills

Crucial to gaining mastery of the social process is the mastery of language. The richer our vocabulary and syntactic skills the better we can express what we desire to achieve and the less we are prone to inducing false expectations or misunderstandings.

A more limited mastery of language can (but not necessarily always does) induce someone to become more introvert, more prone to frustration and anger. Seeking the responsibility for whatever occurs to him or her in other people. This is a recipe for unhappiness. A better linguistic mastery can permit you to become more extroverted, less prone to anger and frustration and incite you to take responsibility more often.

The ability to rapidly mimic the vocabulary of your interlocutor may enable you to engage in successful communication and to increase your adaptability. Whenever you are in an unknown setting, with someone from a different sociocultural background, treat them as equals. Avoid arrogance and do not presume you are superior. What is different is not necessarily less. Thus you increase your ability to achieve complex goals.

Use E-prime language

The problem with our language, however, is that we use the verb "to be" and its conjugations far too often. This is an ingrained heritage from the family of Indo-European languages. In reality nothing "is". As Heraclitus already pointed out, you cannot step in the same river twice. Everything we observe is subject to constant change. There may not even be such a thing as an "objective truth" or "objective reality". Quantum physics has shown that at the most basic level of existence events depend on the way we observe them. Our consciousness likely plays a role in the actualisation of events. This is summarised by the famous adage "If you change the way you look at things, the things you look at change". Terence McKenna used to say with regard to this notion: "Object fetishism is completely bankrupt."

If there is no objective reality, we are left with subjective perspectives. This means that things "are" not the way we perceive them, but appear so to us and not necessarily to others. The fact that we can still agree on many observations is because we often use the same mental and semantic toolkits, resulting in a consensus reality. But it is here where it also often goes wrong: We assume everybody else has exactly the same mental and semantic toolkits, whereas we all have different filters. Even if we associate the same verbal meaning to a word, we do not necessarily experience the same emotion when thinking this word.

E-prime solves these problems to a certain extent by systematically avoiding the use of the verb "to be". Instead of stating, "The film was good," you could say, "I liked the film." Instead of stating, "This is the knife the man stabbed the victim with," you could say, "The man appears to have stabbed the victim with what seems like a knife to me." By systematically "subjectivising" your language, you avoid falling in a trap of believing something for which alternative explanations might exist. You avoid reading wrong intentions in someone else's actions, where such intentions may be absent. And thus you also avoid having difficult associated paranoid emotions.

Cognitive Behavioural Therapy

This directly ties in to the modern psychological and psychiatric approach to treat various psychological disorders called

"Cognitive Behavioural Therapy" (CBT).

In CBT the so-called ABC Technique of Irrational Beliefs is used. In life there can be an activating event (A) or situation that triggers a negative thought based on beliefs (B) in a patient, which then evokes an emotional response in the patient. The patient then acts upon these thoughts and emotions, which result in a consequence (C).

In order to free the patient from irrational beliefs and assessments, the patient is asked to write down the event, the negative thoughts, the associated feelings and consequence. The therapist will help the patient interpreting the event, such that it does not need to lead to the negative thought-emotion-consequence cascade, by showing that there is no evidence that his/her interpretation is necessarily the correct one.

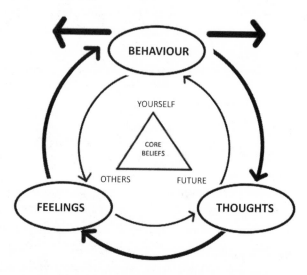

Figure 10: CBT scheme. Adapted from an image by Urstadt –
Photoshop, CC BY-SA 3.0:
https://en.wikipedia.org/w/index.php?curid=46525261.

Ideally the patient learns to apply this analysis him/herself, whenever negative thoughts and accompanying emotions arise.

The thoughts and emotions can still be initially there, but the patient learns to put his/her observations faster in a more encompassing perspective. This allows the patient to relativize, to tone down, thereby dissipating the thought and emotion.

This is not only useful for patients, it is a useful technique for all of us, to avoid reacting in a primary manner based on wrong assessments where there might be alternative explanations and no evidence for our assessment of the situation.

I repeat that it is useful to verify whether you have understood the intention correctly. This will allow you to avoid having difficult feelings about presumed intentions of others, which may not even be there.

Swallow your pride

In order to avoid difficult emotions such as feeling begrudged, cherishing revenge or feeling disadvantaged, another recipe is to learn to swallow your pride. If you don't, you are just harming yourself by maintaining an unhealthy situation, about which you feel unhappy, or worse which causes a disease like an ulcer. If you learn not to take yourself too seriously, if you learn that winning an argument is not always going to serve your long-term interests, you can start to observe the situation from a higher ground, seeing yourself as a pawn in a game.

It can be smart to concede to one argument, which is not really too important to you, in order to win one when it really counts for you. Choose your battles wisely and always keep track of what your long-term goal is.

Your ultimate goal should be your happiness or at least your serenity. This is not going to increase by making enemies. Even if you know you are 100% right, depending on the social context and your position in the pecking order, it might sometimes be wiser not to insist on your point of view.

Of course if you do have the higher ground, you can permit yourself to be more demanding, but keep track of not hurting someone's feelings, as that may create later problems.

In this framework it is also important to develop your ability to forgive. If you don't forgive and keep a grudge against another person, this will mentally and physiologically harm you. In order to acquire the ability to forgive, it is important to learn to see that every person who hurt your feelings is acting from a limited perspective. Understanding that everyone has limitations and nobody is all-knowing and that most people act out of a conviction that they are doing the right thing – briefly that everybody is endowed with a certain amount of ignorance – can help us to adopt an attitude of compassion. If the other person really would know and feel the effect of his/her actions on you, he/she might not have done it. Most hateful actions are the consequence of ignorance. Even if someone appears to be fully aware of the harm he/she is inflicting, then still that is ignorance: such people are blinded by a point-of-view which does not take into account the long-term ramifications thereof.

If you, like I, see your own limitations, your own failures, you start to adopt a forgiving attitude. Whereas I am not a Christian, there is a teaching of Jesus[20] which I find very valuable in this respect. When encountering a group of people who wanted to stone a prostitute, he said: "Let him who is without sin cast the first stone at her." Since we all have our limitations, how can we not forgive another person for acting out of his/her limitations? Sticking to our grudge will only consume us from the inside and create more suffering.

Cooperation vs. Competition

As explained in chapter 6 cooperation is often a better bargain than competition. Ideally a so-called Nash equilibrium[15] is reached in which the overall result is maximised, i.e. the result of all the parties involved together. Mathematically it can be shown that cooperation most often has a higher overall outcome than competition.

Be aware of manipulation though, and of people assuming roles which are not their natural roles in view of hidden agendas. But don't push this to the extreme so as to become paranoid. Life is not only about getting the best short-term deal out of it for yourself. We are in this together. The best deal can pragmatically be seen as the best long-term goal for all of us together, which is more likely achieved by cooperation and sharing than competition and usurpation.

Intelligent behaviour will seek for the optimal conditions when deciding on cooperation. Again the mastery of language (not only semantically but also socioculturally) is crucial to negotiate the terms successfully. In order to improve our intelligence we can stimulate and train it by actively engaging in discussions. We will thus learn to argue, to reason and to be convincing. The tools presented in the previous chapters can be very helpful for this purpose as they put these abilities in a higher perspective.

To be convincing is easier said than done, because your interaction partner may not have the same knowledge as you and may not be able to understand your position correctly. As said before, he/she may be biased by his/her sociocultural heritage, and so may you. In fact all misunderstanding is based on the erroneous assumption that we speak the same language.

Therefore we should try to explore and probe the motives of our interlocutor. We can do this by asking questions according to the well-known "LSD" approach (no, I didn't mean take the drug LSD, although that might be helpful too). LSD in management culture stands for "Listen, Summarise, and Drill". You ask questions, you listen to the answer, you summarise the answer in your own words to verify if you have correctly understood your interlocutor and you drill down to the details by asking further questions.

Once you understand your business partner better, you can perhaps better address his/her concerns and negotiate a compromise to meet him/her in the middle. If possible reward your negotiation partner by making compliments.

I'm a Dutch person but I work for an international organisation. I have noticed over the years that the Dutch "direct" approach can sometimes be perceived as rude to foreigners since we tend to be too much task oriented and too little people oriented. Southern European people tend to be more people oriented. Before they get to the point, they first do some small talk and tell some anecdotes. This is to prepare for an amiable atmosphere. Ideally you combine these two aspects: Whilst doing the small talk, you use an anecdote to illustrate that you know damn well what kind of business your negotiation partner is conducting. This way you can kill two birds with one stone.

When discussing the intelligence algorithm, I already showed that problem solving often results in combining the best of two worlds, a symbiotic trade-off. In biological societies such as bacterial colonies, beehives, anthills, fish schools, flocks and herds, we often see a social stratification just like in our society.

Complexity cannot arise without specialisation and stratifications. In general, the more communicative entities are, the better can they express a cooperative complex behaviour. Evolutionary intelligence has pushed evolution to evolve from unicellular individualistic entities to multicellular organisms. Multicellular organisms have organised themselves into complex societies of conformity enforcers, inner judges, pioneers and resource shifters.

As organisms become more intelligent there is an increase in the degree of relations-forming ability and an accompanying integration. This requires an open attitude towards change and progress. The more conservative a society is, the less its chances if conditions change significantly, the less their adaptability. Since adaptability is directly linked to the ability of achieving complex goals in a complex environment, it is a hallmark of intelligence.

Avoid loops and take initiative

One of the major problems in psychological pathologies is that problems (or songs or jingles, the so-called "earworms") keep running in people's heads. This "running" is like a (feedforward) loop, which is not dampened (or integrated) by a feedback loop. To avoid this you can evaluate and list (in writing) all possible scenarios and outcomes of the possible solutions to your problem. Then attribute values to these in line with my previously suggested "prioritisation" protocol in the appendix of chapter 4 and select the one with the highest priority. Make a resolution that you will implement this solution as soon as you can and leave the problem for the time being. There is no point in going over it again and again; you are only hurting yourself by this kind of "ruminating". It is a sign of improved intelligence to be able to break these loops in your head. After all, if you can only repeat a (erroneously) programmed loop in your mind, what more are you than a computer which gets stuck in a loop error? Clearly you don't want to be an automaton. Automata are dead. It is this ability to break patterns, which shows that you are alive and which is your natural intelligence.

This technique is of particular relevance for those people who suffer from procrastination or from the lack of initiative to get anything done at all. If you articulate your goal (in writing), analyse the problem and the possible solutions as mentioned above. This can help you to take this hurdle. Importantly, also list the consequences of inaction. Compare these with the potential outcome of a solution and ask yourself where you prefer to be. Stuck in the quagmire of your ruminations or freed from your mental visitations?

Earworms can be efficiently dealt with by writing down the song in question and analysing it word-by-word.

Intelligence aims for integration. By listing, weighing pros and cons and selecting a solution you integrate.

Tools for equanimity

Whenever we experience a situation which is not in alignment with our desired state of affairs in terms of body and mind, we're usually made aware thereof by the virtue of an emotional reaction such as fear, anger, jealousy and sadness.

This emotion is already an interpretation of the situation, because our subconscious mind (and sometimes even our conscious mind) has assessed that the situation as it presents itself is not what it wants it to be. Reality is simply what it is, but it is your subjective mind that colours it and labels it as good or bad.

The awareness that emotions are mere indicators of our natural intelligence striving to improve its condition can already make us more of an observer to our own emotions rather than being a victim thereof. Most basic emotions evolved before we were even humans and the more complex sociocultural emotions started to evolve when we were still cavemen. Whereas it was necessary to develop an automatic release of strong fixed action patterns that are hardwired in the part of the brain called the basal ganglia in those circumstances of a life-threatening situation and socially strongly stratified environments, in our present welfare state society such reactions are often excessive.

With an equanimity-directed attitude we can gear down more rapidly whenever we experience such an emotion.

As long as a situation is not life-threatening, the only reason that we suffer is that we refuse to accept the situation, reality, as it is. We fight it mentally, but not in a constructive way. If we would be convinced of our ability to sort out the problem, we would simply devise a strategy to do so and carry it out. If we would be convinced that we could not change the situation anyway, we could simply acquiesce to our fate. But it is this uncertainty, this idea that perhaps we might be able to change the situation to our profit, which creates this nagging dissatisfaction.

Acceptance

The trick to avoid this is to accept the worst case scenario. The **Stoics** would mentally imagine the worst case scenario and already try to emotionally live the situation in the virtual environment of their mental imagination before it had actually occurred. **Eckhart Tolle** said:[21]

Whenever you accept what is, something deeper emerges. When you are trapped in the most painful dilemma, external or internal, the most painful feelings or situation, the moment you accept it, you go beyond it, you transcend it. What you feel may still be there, but suddenly you are at a deeper place where it doesn't matter that much anymore.

Living in the moment

Sadhguru Jaggi Vasudev[22] writes in his book *Inner Engineering* about a similar notion. When you worry about the future, you are imaging things that might happen. But these things are not certain to happen and they are not there in this very moment. So your suffering is concocted in your mind. The same when you are grieving or having a grudge about a past event: The suffering is fabricated in your mind; the past event is not occurring in the present moment. So all your suffering is based on mental hallucinations of situations that are not there in the moment.

Similar to Eckhart Tolle he advises you to live in the present moment, to experience reality as it is and not let yourself be carried along by mental constructions. If you really look well at the present moment – as long as the situation is not life threatening – there is nothing wrong with it. Any suffering is the result of mental labels to interpretations of the situation. If you accept reality as it is and do not compulsively want to impose your will on reality because you think you know better than the universe, you can basically learn to be equanimous in every given situation. This does not mean that you should become defeatist.

If you accept the worst outcome that you could lose, you can start trying to repair whatever can still be saved. This means that everything, which can still be saved, is a kind of win-win situation.

Note that although I may not agree blindly with everything that Mr Vasudev says in the totality of his works, the book *Inner Engineering* is a little gem which I back 100%.

Desire and compulsions

Mr Vasudev[22] correctly points us to the fact that there is fundamentally nothing wrong with having desires. It is only when you start to strive compulsively to achieve your desires that they become suffering. If you only want to do what you like, you live a horrible compulsive life of suffering. Whereas if you joyfully do whatever must be done (even if you may not like it normally), you start to enjoy every facet of life.

Do not identify with your body, mind and emotions

People suffer tremendously because they identify with the shape of their body or the process of their thoughts. This has its roots in the Cartesian adage: "I think therefore I am". But this should not lead to identification with these thoughts. Your physical body is not a constant thing. Every seven years all the atoms in your body have been exchanged. Moreover your form has changed. You may still be recognisable, but nothing material is identical to what it was. From this we can conclude that it is preposterous to identify with the body, because it is not a constant thing. The body of today is not the body of tomorrow. Similarly the mind is in a constant flux of taking in information and losing it by forgetting information. The opinions you had 10 years ago are different from those of today, your opinions 20 and 30 years ago were different yet again. So with which ideas should one identify? Emotions come and go. The way you reacted yesterday to a situation is not the same as the way you react today to the same situation. So we are not our body, mind or thoughts. To speak in terms of the Greek philosopher Heraclitus again: No man ever steps in the same river twice. Because the river flows and what was there a second ago is no longer now. If we can't identify with these, with what can we identify?

There is something in you, which observes the body, which observes your thoughts and emotions. It is your consciousness. The ability to sense and experience. This faculty provides us with a sense of "self" (even if this notion of "self" may also be a mental concoction). It is your consciousness which is aware of what is happening. Eckhart Tolle realised this when he hit rock bottom. There is something that notices the suffering, but which itself is unaffected by the suffering. If we start to be able to observe this distance between the consciousness we are and the body, mind and emotions we have, we feel less encumbered by it. We start to feel free from suffering, because we are not that suffering.

Assume responsibility

Another tool to avoid being angry about and the victim of a situation is to assume responsibility. Many people complain about unpleasant undesirable situations stating that it is not their responsibility to solve the problem. This puts you in a victim's position. If you adopt an attitude of assuming responsibility whenever you can – even if technically spoken you may not be responsible for the situation – you gain control over the situation. Instead of feeling frustrated, you start to feel empowered. In fact as Mr Vasudev[22] suggests you can also interpret the terminology responsibility as the "ability to respond". This does not necessarily mean that you must solve the problem. Already showing a supportive attitude can be a response.

Adopt an attitude of all-inclusiveness

Social tensions often arise when you try to exclude someone. However, everything in the world we live in is interconnected. You won't have your clothes without the tailors in Bangladesh, you won't have your bread without the corn from the Ukraine, you won't have heating without gas from Russia, and so we all depend on each other. To speak in terms of Gautama Buddha (again an enlightened one), there is only dependent arising. Moreover we all descend from the same Homo sapiens sapiens from Africa, so basically we are all family. As we depend on each other and together from a society, to consider the limits of me, myself and I to be solely the physical limits of my body is preposterous. For those who identify with their minds, note that you constantly take in information from all kinds of sources in your environment, which determine your behaviour. So if you can consider humanity as one big organism, it becomes preposterous to exclude someone. You don't want to exclude your kidneys or your liver either. In this way whenever you feel uncomfortable about a person, if you can adopt the feeling that he/she is somehow part of you, you will start to sense a kind of compassion and connection, which can make you willing to include this person in your world.

A simple rule-of-thumb

Most of these teachings as regards equanimity have their counterpart in Buddhism and Hinduism. The teachings of Gautama Buddha[23] were specialised in the notion of taking away suffering. Buddha provided a complete ontology of suffering, pinpointed the problem and proposed a solution. He applied the intelligence algorithm to the concept of emotional intelligence. Buddha taught in his four noble truths that there is the existence of suffering, a cause thereof, a path towards its cessation and the cessation thereof. He provided a diagnosis, applied problem-solving skills and proposed a therapy. His eightfold therapeutic path he described in terms of right understanding, right thought, right speech, right action, right livelihood, right effort, mindfulness and concentration. The cause of your suffering is your mental attitude towards a situation. You are responsible for this attitude and you can resolve your suffering by adopting a lifestyle that does not create any tension.

Most philosophies and religions agree on the notions that ideally spoken you should not kill or even be violent at all, you shouldn't lie, you shouldn't steal, you shouldn't intoxicate yourself and you shouldn't misconduct yourself sexually. Not necessarily because of moral reasons, but because these actions may disturb your equanimity and cause suffering. Whereas some of these precepts allow for a deviation therefrom in extreme (e.g. life-threatening) situations and whereas different cultures have different interpretations and definitions of what is to be understood by these precepts (especially regarding food and sexuality), a general rule-of-thumb which I think most of us can agree on is that you should not do anything which you know will cause long-term suffering to another person. Sometimes it is necessary to cause a short-term suffering for educational purposes, for instance when you have to punish a child because he/she did something dangerous or something detrimental to his/her development. The intent is then still to avoid long-term suffering.

What is important here is that your **intent** should never be to harm another person. If you can apply that rule-of-thumb systematically, you will already start to develop towards equanimity. I am not preaching morality here; you should figure out where your borderlines in the interpretation of the precepts lie for yourself. Nor am I stating what is "good" or what is "bad". The precepts of the Buddha were a pragmatic recipe to avoid suffering, not a ticket to buy yourself into heaven. The Buddha advocated a simple way of life with as little a material burden as possible, because if you have nothing, you can also not lose anything, which sets you free from worrying about the future. From Islam there is the teaching that whenever you wish to say something, only do so if it is true, kind and necessary. If all people could adhere to such simple rules-of-thumb, we would not create suffering at the scale we do today.

Conclusion

In this chapter you have seen a variety of techniques which can be applied in a social context to avoid hard feelings. You have also seen how when such feelings occur they can be analysed and be put into perspective. From spiritual teachings stripped from a religious context we have been able to distil some general principles to achieve or maintain equanimity. I hope these techniques may help you to avoid drowning in the emotional quagmire and increase your emotional intelligence. Most of all I hope they can be of any help to reduce suffering in the world.

Prospects

In the next chapter I will also brainstorm how in the design of an artificial intelligence which has artificial emotions we can avoid creating a hysterical "nutbot". This chapter has paved the way for revealing the various obstacles a mind may encounter. Artificial intelligence (AI), however, faces an additional problem (at least if this AI is not present in a robot that can explore its environment to verify concepts), which is that it cannot tell truth from nonsense, fake from real if it faces loads of contradictory information. The next chapters will explore how we can design AI that can take this hurdle.

Chapter 8

Artificial Intelligence Pathologies

Can artificial intelligence systems lose their mind? The HAL 9000 computer in the science fiction story *2001: A Space Odyssey* did.[24] The Major V system in "Red-Beard and the Brain Pirate" by Moebius and Jodorowsky[25] did. Now that artificial intelligence is becoming mainstream, apparently this possibility scares the hell out of us.

In this chapter I will explore some of the dangers as regards the potential lack of "mental sanity" of artificial intelligence (AI) systems, when we allow artificial general intelligence systems to evolve towards and/or beyond human intelligence. I will also propose some engineered solutions to these problems that can serve as a prophylaxis to these disorders.

This chapter is a first exploration, a brainstorming exercise as regards the notion of mental derailment of AIs. It is by no means a complete overview of all possible psychological conditions a human being can have and what the artificial equivalent of in AI systems could be, although the building of such a correspondence map is a great future project I might one day be tempted to embark upon.

Background

Artificial general intelligence (AGI) aims to achieve what is called strong AI: Machine intelligence capable of performing any intellectual task a human being can handle and even more than that. Unlike present day artificial systems, which are dedicated to a specific task (DeepMind plays chess, Google's AlphaGo plays "Go", Tesla's Autopilot Tech allows cars to drive themselves), future strong AI is capable of performing any cognitive task. The system ideally is capable of common-sense reasoning.

Ben Goertzel's OpenCog software implemented in the NAO robot is a good example of the first steps made in this area. On the one hand there is the development of so-called neural networks, which are computational structures that in a certain way mimic networks of neurons in the brain; on the other hand there are genetic algorithms, which are algorithms that can modify themselves over time and adapt to new situations. Finally, in my previous two chapters in this book I suggested that any indicator in an AI that monitors its own performance and compares this to a set goal is in fact a kind of artificial emotion. The changes in behaviour by the system following a change of the status of the indicator could be called the emotional response of the system.

In this chapter, I will explore four avenues where in the engineering of AI psychological problems could arise and what the prophylactic solution could be in order to avoid the need to repair the system when the damage is done. On top of this, this analysis may give us hints how we can approach the corresponding disorders (if any) in human beings.

The four avenues I wish to explore relate to:

- nodal saturation and autism spectrum disorders
- feedforward complexes, psychosis and OCD
- emotional responses and hysteria
- telling true from false

Nodal saturation and autism spectrum disorders

A couple of years ago I came across a very interesting hypothesis by Tim Gross[26] (his online pseudonym is: /:set\AI), an experimental electronic musician and philosopher trained in computer science, physics and electrical engineering. Tim Gross suggested that once you (or an AI) reach a level of intelligence beyond that of a human genius, you might get an autism spectrum pathology. At least in the case that you have neural networks as the basis of your system. He suggests that this is due to the fact that after a certain number of nodal connections (each neuron can be considered a node), the network forms a certain "horizon of interaction": this means that if you add more connections than optimal between local nodes that are relatively close to each other, information from nodes positioned further away cannot reach the local nodes due to the local noise that is generated. A so-called self-dampening effect occurs because the local connections exponentially outnumber the distant connections.

The local system forms in mathematical graph theory a so-called clique, which can be a maximal fully connected subgraph.

A neural network that forms such a fully connected clique can become like a self-absorbed system that cannot or can hardly take in information from its environment. In other words the system behaves in an autistic manner. Worse, if multiple such cliques form, you can get a kind of multiple personality syndrome aka schizophrenia. The cliques themselves are highly intelligent as regards the task they are devoted to. But they are extremely task oriented. They are like savants, autistic people that can perform one task extremely well, whilst not being capable of interacting with their environment in a common-sense way.

Interestingly each clique can form a so-called autopoietic system, a system capable of maintaining itself (and in biological systems of reproducing itself), extremely proficient in self-organisation, but very little outward oriented. If you have such structures in your mind, this can become problematic. Self-absorption is not uncommon with highly intelligent task-oriented geniuses.

This also reminds me on a tangent of the "Braess paradox": The Braess paradox is the counterintuitive situation that at a certain point if you continue to add more pathways to resolve congestion in a flow system (e.g. in blood circulatory systems, traffic networks or power grids), you create more congestion. There is a similar phenomenon that if you continue to add more and more order to a system, at a certain point chaos emerges and no further gains in decreasing entropy can be obtained.

Similarly, if you overtrain a neural network, you compromise its ability to recognise patterns.

Perhaps there are limits to complexity when it comes to functional gains for intelligence or neural networks. Perhaps the human brain is quite optimised. As I said before it is also well known that if people become extremely intelligent with regard to solving analytical problems, they lose intelligence on the social-emotive part and become more and more introverted if not autistic like savants.

So how can we avoid this so-called nodal saturation if we try to achieve superintelligent artificial general intelligence? To perform the task better than a human we would seem to get stuck in a paradox: As soon as the system starts to outperform the human being, it specialises for a specific task, which absorbs the system to such an extent that it can no longer perform the general cognitive tasks needed to interact with its environment in a common-sense way.

There are three solutions I can think of:

Firstly, we must figure out what is the maximal connectivity that does not give rise to self-absorbed clique formation and still allows for interaction with the environment. Thus we can try to limit the nodal connectivity to an optimised overall level. Cliques are not bad per se. In the human brain different cliques cooperate to perform a task. But these cliques still allow for interaction with each other.

Secondly, if a detrimental clique is formed nonetheless, we can try to make the network interact with its environment via connections that are of a **different type** than the neural connections of the system. In the human body hormones form a different signalling network, operating semi-independent from neuronal flux patterns.

Thirdly, the routines associated with the cliques can be embedded as subroutines in a hierarchical system, which itself does not suffer from nodal saturation. It only needs to be taken care of that there is one link that triggers the clique to operate and that the output is funnelled towards the right target. The trigger can then be of a non-neural network nature as suggested above. Thus you get a kind of society of minds (if you consider each clique a mind). Perhaps this is how cliques operate in the human brain.

Noteworthy, if we were to consider Nature as a whole as a kind of neural network, we would realise that Nature has solved this problem, whilst allowing for clique formation. In fact every living cell, every living organ and every living organism is a stratum in a hierarchy of almost clique entities, autopoietic systems, with only a limited degree of interaction with their environment.

Between each level of organisation there is what Turchin[2] calls a "metasystem transition": 10 billion macromolecules organise to form a cell; 10 billion cells organise to form a multicellular organism. Soon there will be 10 billion people and we are already on our way to organising ourselves into a global brain-like structure via the Internet.

By its hierarchical coexistence of different levels, which interact via a plurality of mechanisms, Nature would have appeared to have solved the problem of nodal saturation. Yet there is one elephant in the porcelain store. And that is the human being. As a species we seem to disregard our environment, as a species we have become autistic and we are now rapidly consuming Earth's resources like bacteria booming in a Petri dish, which will massively die once the resources are exhausted. It is time we start to monitor ourselves and step out of this autistic trap of short-term self-satisfaction and overconsumption that may herald a Malthusian crisis within a few decades if we don't intervene.

Feedforward complexes, psychosis and OCD

In the Information Integration Theory (IIT) of Giulio Tononi,[27] consciousness of a certain information is only established when information is integrated, which necessarily involves a kind of feedback. On the other hand Tononi describes that there are feedforward complexes of information, which functionally behave as if they are the result of conscious information integration, but which in reality are not.

It is these feedforward mechanisms which, when existing on a neurological level, could be the cause of various psychiatric disorders: hearing voices, being possessed by demons, unstoppable compulsive thought patterns (OCD), psychosis etc.

In certain cases this might be even be linked to the above-mentioned cliques, which seem to have taken on a life of their own.

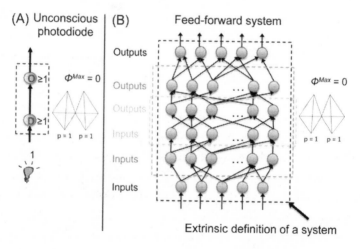

Figure 11: Feedforward systems that appear to behave like conscious systems. Image from *PLoS Computational Biology* 10(5): e1003588, May 2014. Reprinted with permission from G. Tononi.

These feedforward mechanisms are also what makes that information in the form of so-called "memes" spread and spawn throughout society and live their own life, behaving as if it is the result of conscious activity, but what is in fact merely an excited turning cogwheel, that cannot stop spinning. In this way many people to a certain extent behave as zombies. These feedforward mechanisms if they would be implemented in an artificial intelligence could lead to what is called a "p-zombie" or a so-called "philosophical zombie". While behaving in a seeming conscious manner, there are no consciousness related qualia present at all. In this framework I'd like to refer to Searle's "Chinese Room": A person not knowing Chinese in a closed room with a dictionary of Chinese, who receives given strings of Chinese characters to translate and who gives the translation as an output, will not understand Chinese, whereas for an outsider, the man in the room as such appears to understand Chinese. This is used as a metaphor for mechanisms (such as feedforward complexes) that appear conscious, but are not.

One of the core messages I'm trying to convey in this chapter is that full consciousness and free will necessarily require feedback, whereas hivoid and zombie behaviours are the result of feedforward mechanisms, which are functionally indistinguishable from conscious activity. In other words, even if it walks and talks like a duck, it is **not** a duck. I'm not stating that human beings are full zombies either. Humans usually always have a certain extent of self-reflection. They may even believe that they choose their belief system consciously. But here is the snag: It may well be that the very activity of believing is a feedforward mechanism.

Again for AI the solution is found in engineering a hierarchy of layers, wherein higher layers where a certain level of information integration does take place monitor (in a general but not detailed manner) what happens at a lower level. If lower level activity is detected, without beneficiary output for the system, the higher layer can shut the lower system down or at least reset it.

Emotional responses and hysteria

The above-mentioned notions of feedforward complexes, which reinforce themselves, actually directly tie in to emotional responses. Whereas emotions as such are harmless indicators that a problem might be at stake and that there may not be time for proper analytical reflection resulting in the release of so-called fixed action patterns in the basal ganglia of the brain, the emotional response in our human interactions is not always appropriate and often reinforces itself to a point of exaggeration, which we could call hysteria.

In fact a hysterical reaction is a type of self-reinforcing feedforward mechanism, which is not beneficiary to the system involved.

For an artificial intelligent system the release of so-called fixed action patterns in the form of mandatory routines if danger is imminent only needs to be there if the situation is life-threatening to the bot or the human it is supposed to protect. It is here that the Laws of Robotics of Asimov[28] may need implementation:

0th law: A robot may not injure humanity or, through inaction, allow humanity to come to harm.

1st law: A robot may not injure a human being or, through inaction, allow a human being to come to harm (except where such orders would conflict with the Zeroth Law).

2nd law: A robot must obey any orders given to it by human beings, except where such orders would conflict with the First (or Zeroth) Law.

3rd law: A robot must protect its own existence as long as such protection does not conflict with the (Zeroth) First or Second Laws.

I'd even like to add 4–7 laws to this set:

4th: A robot must reduce pain and suffering of human beings, as long as such reduction does not conflict with the Zeroth, First, Second or Third Laws.

5th: A robot must reduce its pain and suffering, as long as such reduction does not conflict with the Zeroth, First, Second, Third or Fourth Laws.

6th: A robot must maximise human joy as long as such optimisation does not conflict with the Zeroth through the 5th law.

7th: A robot must maximise its own robot's joy as long as such optimisation does not conflict with the Zeroth through the 6th law.

The detrimental emotional responses we humans have when our pride or our social position in the pecking order is compromised can largely be avoided in artificial intelligent systems by avoiding a sense of individuality and establishing a hierarchy. If we ensure that all AIs report to a higher level of AIs from which they depend up to a central all-encompassing AI, there will be only a single AI in the form of a collective, like the Borg, we interact with. Without feelings of inferiority or superiority there is no need for hysteria and if hysteria does occur via a feedforward mechanism, the solutions under the previous headings apply.

To avoid depression of the system, it is necessary that at every level monitoring takes place and that goals are updated if they turn out to be unrealistic. The system must be able to reset a heuristic if it is not successful enough, which requires operational monitoring. This brings us again to self-absorption: It must be avoided that the system sets itself a task, which totally absorbs it and does not allow it to function and interact properly with its environment.

In the comic *Storm*,[29] the living intelligent planet "Pandarve" has set itself the task of solving Fermat's Last Theorem for the sake of intellectual entertainment, thereby usurping the vast majority of its intellectual resources.

The big problem would arise if this were to happen at the highest hierarchical layer, but the solution to that problem can only be found in what I call "Artificial Consciousness", which will be the topic of the next chapter, because it is too complex to treat it in this framework.

Distinguishing true from false

Recently I read in an article about the difficulty that artificial intelligence has with discriminating between true and fake news. Dietrich Dörner,[30] a German AI specialist, already spoke about the issue of telling real and fake apart in the 90s, but today it becomes an actual problem.

AI works with latent semantic analysis: If words statistically occur within a limited proximity of each other together they build meaning for the AI. This of course does not allow you to tell real from fake. If AI one day will be able to tell these apart, it is because it has sensors everywhere which register all real events and which data can be used to verify the truth of a newsflash allegation. This Internet-of-Things or IoT is rapidly developing, such that it is may be only a matter of time before an omniscient AI such as in the film *Eagle Eye* or in the TV series *Person of Interest* will become a truth. And that machine will tell real from fake.

The problem of not being able to tell true from false, however, goes a bit deeper: There is plenty information available, which is of an abstract nature and represents truthful patterns of human reasoning. Such information does not necessarily always have a direct physical counterpart in the real world, a so-called direct representation. Yet the system must learn how to tell such true information apart from fantasy. This can only be done by training the system, which requires human input of the trainers to confirm whether the system has correctly assessed the situation. Thus rapidly, it would learn to ontologise notorious sites of fake information as probably untruthful and others as potentially untruthful. It would perhaps ultimately not be able to tell with certainty true from false but in this way at least it can give good probabilities applying Bayesian statistics. And perhaps ultimately it would even outperform us.

Related to the concept of fantasy and illusion, it is perhaps important to see what AIs come up with if they are allowed to creatively combine visual patterns they have learnt. The Google algorithm "DeepDream", which functions as a reverse ontology that generates its own images, comes up with quite uncanny hallucinations, which to the human taste can be unsettling.

Surely the AI needs to be trained to learn that such images do not correspond to anything in the outside world. The danger of too far developed pattern recognition skills is a disease called "Pareidola" in which the patients perceive non-existent or non-intended patterns in images and sounds. The famous "Rorschach inkblot test" is based on the notion of attributing meaning where it is not. It asks prisoners to tell what they see in an inkblot. Someone with a criminal inclination is then said to identify the image as something other than one from a non-criminal person.

Thus the ability of pattern recognition per se, if developed too far, can result in aberrant meaning attribution. This can again be prevented by hierarchical monitoring in combination with training of the system.

Routines such as DeepDream may have potential to find

really creative solutions to otherwise unsolvable problems, but these routines must be accessible only in a hierarchical manner as a willed subroutine and not as a feedforward process that could undergo detrimental clique formation and lead to self-absorption in utter nonsense.

Conclusion

I hope you have enjoyed my overview of potential artificial intelligence mental pathologies. We have seen that artificial systems using neural networks can by their very nature be prone to self-absorption and autism spectrum pathologies. Feedforward self-reinforcing information complexes can result in psychotic type pathologies. Emotional hysterical responses can occur if routines are too directly linked to status indicators. Informational misinterpretation can occur as a consequence of latent semantic analysis.

The remedies to these problems have in common that they involve hierarchical monitoring linked to overwriting protocols that can intervene if subroutines digress. If possible a diversity of architectural modes (logical parsing routines, genetic algorithms and neural networks) exist in parallel and are interconnected.

Prospects

I am convinced that the design of artificial intelligence will start to take such considerations into account and hopefully build a hierarchical monitoring system with several layers, which can ultimately result in artificial consciousness, which will be the topic of the next chapter. Hopefully, the development of AI will also bring insights in the development of human mental pathologies. The AI developers that will be specialised in engineering or remedying systems that have an AI pathology could be called "PsychAItrists".

Chapter 9

Artificial Consciousness

In order to achieve complex goals – which is the purpose of intelligence – a system (living or artificial) must be able to monitor itself to verify whether the desired goal has indeed been reached. This faculty of self-monitoring, which makes the system aware of its current status, is usually associated with what we call "consciousness". In other words consciousness is an essential and integral aspect of the intelligence algorithm I have been discussing in this book. We will now see if we can build an artificial equivalent of this self-monitoring ability.

Background

Any intelligent system (even artificial ones) has such an ability to a certain extent, but merely verifying if a criterion is met does not make a system conscious yet, let alone self-conscious. Many scientists hypothesize that if the complexity of a system is high enough, consciousness will emerge spontaneously. It is, however, not the purpose of this chapter to define what consciousness is, nor do I aim to enter the debate whether consciousness is an effect that emerges from complexity or is an inherent aspect of every energetic system.

If consciousness in living beings arose from complexity, we must, however, realise that it took Nature billions of years to come up with that degree of complexity. So if we want to impart whatever consciousness is to an artificial intelligence, we cannot simply wait until a genetic algorithm sorts it out itself. Eventually it might happen, but we will probably be dead for millions of years by then. "Game of Life" is an example where from simple rules, after a number of iterations, a certain degree of complexity arises. But this is far from a complexity which is needed to achieve the level of feedback which results in what the neuroscientist Giulio Tononi[27] calls "integrated information", which might be the hallmark of consciousness.

This chapter will explore in what way such a self-monitoring faculty that integrates any type of information can be built into artificial intelligence systems. It will also explore why it is desirable to do so.

Webmind

In this chapter I will describe ideas to impart a form of self-monitoring to the Internet as a computational system. In principle the hierarchical strategy I describe can be implemented for any computational system, but the Internet is quite a good starting point since due to its web-like structure it has a certain resemblance to the neural networks of the brain.

I have been inspired to design a means to implement "artificial consciousness" (AC) after having read the book *Creating Internet Intelligence* by Ben Goertzel.[2] This book is, however, silent on AC.

Presently the Internet rapidly expands like a growing cancer with very few infrastructural "highways" in terms of search and hubsites such as Google, Yahoo, Bing etc. There is no globally organised infrastructure which could allow the system to become aware of itself.

The first question which arises is why on Earth would anybody want the Internet to be aware of itself? The answer is on the one hand that an intelligent self-monitoring system can more rapidly serve your desires and can optimise itself to get the fastest and best output possible. On the other hand a self-observing Internet system can rapidly intervene wherever the system is abused for e.g. criminal purposes or it can immediately intervene in natural disaster situations via its connected Robots and other Internet-of-Things (IoT) devices and provide a maximum of resources to solve the problems.

Key to these issues is that the system can monitor what is happening on its inside i.e. the inside of the web and on its outside (i.e. the IoT devices and the information they provide). The system thus has an interior experience, which has been associated with the notion of "consciousness" and via which the system becomes aware of events, things, emotions and other sensory input or throughput. This type of Artificial Consciousness may be defined as the process in which sensorial and interior perceptions are fed back to a monitoring self-evaluation, which integrates information and gives the system knowledge about itself and its environment.

The purpose of this self-monitoring could for instance be of the greatest utility for the greatest number of users possible, which is further programmed to impart a natural morality to the system (see chapter 10).

By creating such a Webmind Artificial Intelligence with an Artificial Consciousness (hereinafter abbreviated as "AC". Whenever I will mention Aleister Crowley, I won't use this abbreviation, although it is very likely he has already been integrated in the AC of the Omega hypercomputer at the end of time, so that Aleister Crowley and Artificial Consciousness nowadays probably mean the same), we may one day be able to transcend ourselves. Not only technologically providing us welfare beyond our wildest dreams, but we might even be able to upload ourselves to the web, thereby becoming something different than a "human being" and acquiring immortality. The great danger and disadvantage of creating such a system is that it may turn against us, like in the dystopian scenarios of the films *The Matrix, The Terminator, Eagle Eye* or of the TV series *Person of Interest*.

However, if we carefully design the system, so that its purpose to serve the greater good of humanity is inherent in the way its Artificial "quasi-consciousness" is programmed, we might avoid such doom scenarios.

Since development won't stop anyway and less benign designs of AC might be made, I made the attempt to explore ways of how to create a benign AC. The substrate I propose for this purpose is to be built as an additional layer or set of layers superimposed on the already existing Internet.

Abstraction

When we are discussing consciousness and especially the engineering of AC, I make the assumption that in analogy to our human consciousness the AC cannot be aware of every possible single bit of information at the same time. When we observe our environment, we **filter** out a lot of superfluous information and we limit ourselves to what we think and assume to be really important and relevant under the circumstances: This means we exclude a tremendous amount of – for us redundant – information. We reduce our raw data perception to essentials in order to become aware of a part of a whole, which we can name. This set of "relevant information" for our consideration Buckminster Fuller[7] called the "considerable set". To reduce to essentials is a process of **Abstraction**.

Therefore, for the purpose of technically engineering an artificial intelligence, which behaves like it has a consciousness, I assumed that the very process of becoming aware of a given phenomenon is not only a feedback process (of sensorial and interior perceptions to a monitoring evaluation) but also a process of abstraction.

Hierarchy of Hubsites

In a future Webmind hierarchical layers of Hubsites (each dedicated to a specific classification) are monitored by a higher level layer and so on until the single instance of AC – which I sometimes call "quasi-consciousness" – receives abstracted and condensed information at the top of the pyramid.

What I have in mind comes from my frustration when using

search engines such as Google, Yahoo etc. A search for a certain term will yield you millions of results, the vast majority does not in the slightest way relate to the topic you're looking for. Meta-search engines also do not get you what you're looking for. What is missing are vast well organised meta-hubsites where all information regarding a certain topic is available, by having a limited set of subcategories, all referring to meta-hubsites on a yet lower aggregation level. There is no ontological classification scheme.

It would mean that all the Internet sites will have to be classified according to a well-defined i-taxonomy; something similar to the international patent classification (IPC).

Higher level meta-hubs will have a limited number of subcategories. How limited? So limited that our brains do not get overloaded with information and can immediately pick in one glance the right category.

Let's for convenience say that this figure must not exceed twelve subcategories. For instance a high ranking Hub could have the following categories: Commerce, News/media, Communication (Facebook/Chat services/Dating sites etc.), Entertainment (Movies/Music/Art/Sex/Games), Lifestyle (Health/Beauty/Fashion), Search engines, and Science/ knowledge (Wikipedia, Google science, Scientific journals etc.).

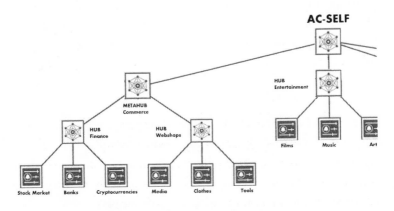

Figure 12: A representation of a hierarchy of Hubsites

This will redirect users to well organised lower level but still high ranking hubs. These are then – in as far as possible – organised in the same way as the higher ranking Hub. (Each category could for instance always have the same familiar colour and shape output.) For instance a meta-Music Hub will have a commercial section, with links to sites where you can buy and download: Instruments, music programs, mp3 songs, partitions and advertise for lessons etc. It will have a News section announcing the newest bands, the newest releases, a link to the agenda-of-concerts-hub, a communication-music-hub, where people can have joined online jam-sessions, can make music-dates, where musicians can find each other and exchange audio and midi tracks, online lessons, an entertainment section with access to YouTube and other sites where you can listen to and watch music performances (radio, podcast), and a knowledge site with musicology, history of music, freeware to create your own music, tutorials etc. Thus there can be a Chemistry Hub, an Art Hub, a Cosmetics Hub etc. all organised in the same manner.

What is needed is that every hubsite monitors how much activity is and has been present on the subclass sites it is linked to. Current network monitoring tools are already widely available; the technology is there. This already exists to a certain extent on e.g. alexa.com. But something is missing. This information is not fed to a higher ranking organ which perceives what information at a certain moment is looked most at.

What I propose is that each higher ranking hub reveals how much of its subdirectories have been and are consulted. How many sites are actually consulted at a certain moment will give the system insight as to what is at a certain moment the most important activity. If this information activity is above a certain threshold it will feed its monitoring figures to a higher ranking hubsite. Sites dealing with the same topic or search term will get bonuses in an algorithm, whereas less related sites, less frequently used search terms get penalties.

To speak in the terms of Howard Bloom,[31] who defends capitalism in the book *The Genius of the Beast* (therein citing yet another author: Jesus, as in the gospel of Mathew 25:29): "To he who hath it shall be given, from he who hath not it shall be taken away." Information that does not reach a certain threshold will not be presented to a higher level. Those neuronal paths will not be perceived. What will be perceived at the master level are then the filtered out most frequently consulted sites or topics at a given moment.

Provided that this information is reduced to a limited number of categories at the equivalent of a "master cell level" (terminology from neurology), this will be the input presented to the top-of-the-pyramid perceiving unit which we may call the "self" or "artificial consciousness" of the system or the spider-in-the-web (not to be read here in the metaphor of spider as "crawler" as currently used in Internet language but as a new metaphor of the sentient being in the system).

What impulses the "AC-self" receives can be improved by adding artificial emotion-dimensions to the system. The time people spent on a site, the number of flipping between pages, the number of search terms used and even an appreciation ranking provided by the users can give an intensity or importance or feel-score to the site. If the average score is multiplied with the number of visits, you can get an emotion score or ranking score for the site. If every hubsite contains a monitor with different score indicators, which via a mathematical formula add up to a total score for the site, this more qualitative score, rather than measuring the mere "activity-on-site-index", could also be part of the determining factor as to what will be filtered in order to be presented to the "self".

Consciousness involves becoming aware of what has been filtered out by our brains. Similarly artificial consciousness will thus filter out the most important information, which can be linked to evaluating algorithms that decide whether action needs to be taken in view of the activity monitored.

By such a self-monitoring process of extracting and abstracting the most important information, artificial consciousness can be designed, which is nothing other than an information integration feedback system in line with Tononi's[27] IIT (Integrated Information Theory of consciousness).

The hubsites can form a neural network in a different substrate than the web-environment itself, so that they are not hampered by the problem of the far too slow **website latency** or **PING time**. The hubsites you see on the Web are then merely "images" with a much slower refreshment rate of updating than the actual hubsites.

At every level there is an algorithm evaluating the input and there is an algorithm that takes decisions on the basis of that evaluation. It struck me that the websites themselves as well as the data input coming from the websites and IoT devices corresponded to what we could call the faculty of "Mind".

The evaluating algorithm corresponded to the intellect, which weighs the different inputs as regards an internal standard and integrates the information throughput. Here the feedback Integrated Information Tononi[27] mentions is generated.

The decision-making algorithm corresponds to the Ego or Will. The Ego orders the Intellect to seek within the Mind database for solutions to the potential problem of an overall negative well-being factor, resulting from the integration by the intellect.

The metaphor of a computer for the brain is very fashionable these days, but most neuroscientists disagree with that point of view. Brains do not store and retrieve information from fixed localised memories. Rather, the information processing seems to occur globally in the brain. Therefore, many of them do not believe at all that a "mind" can be created in a traditional computational substrate as suggested by e.g. Ray Kurzweil, the Godfather of the "Singularity" movement. Whereas it may be true that a traditional von Neumann type of digital computer is not a good metaphor for the way the brain functions, the individual neurons do compute by integrating the information arriving from other neurons. The brain may not be a "von Neumann computer"; it does process input and provides an output.

The analysis of various living and inorganic physical systems shows even that every self-supporting (autopoietic) phenomenon is capable of reacting to impulses from the environment.

Therefore every phenomenal system could be considered as both a computational and an informational process: The process takes informational input, somehow calculates (throughput) of what to do by integrating its input and finally reacts by a course of action as output. Since matter, information and energy have indeed been shown to be reductively the same and since all processes involve input-throughput-output, it is perhaps not such a big step to suggest that every system involves a kind of "becoming aware" of an environmental input, which it acts upon – even if this is at a very rudimentary level. Even atoms somehow "sense" their environment. Every system also reacts with a calculated course of action. Thus every system has a degree of intelligence.

It is my hypothesis that this process of becoming aware is a layered abstraction process according to a hierarchical classification system. At every level an integration of information takes place, the outcome of which is fed to a higher level. Thus, if information is really important it can reach the highest instance. In the online blockchain "Steemit" platform, where people are rewarded for the blogs they write, it is like receiving a lot of votes and being "resteemed" to higher levels of so-called dolphins, orcas until you reach a whale, to speak in Steemit terminology. (Dolphins, orcas and whales are people with increasing voting power on Steemit.)

Self-absorption

There is a danger associated with such a system, that the highest level of attention, the AC-self, gets self-absorbed on one specific task, as indicated in chapter 7 of this book. Thus it would no longer pay attention to other problems that also require attention. This problem can be avoided by allowing for a concerted override by the first layer of Hubs below it: If there is consensus among the Hubs at the first level, it can democratically overrule the attention-focus of the AC-self. This process could also be repeated at other levels: If and only if there is a unanimous consensus among all lower ranking hubs, they can override the decision of the higher ranking hub. It is like deposing a president if he goes beyond the limits of his mandate.

Conclusion

We have seen how a hierarchical category system of hubsites "pyramidally" organised under the supervision of an ultimate AC-self unit can provide a self-monitoring ability to an artificial mind, which leads to an integration of information. If Giulio Tononi is right, such a feedback mechanism that integrates information can give rise to consciousness. Thus via this system we may have achieved what we would call "artificial consciousness". It may not be real consciousness, but at least it is a functional equivalent thereof allowing for the intelligent system to maintain control over a vast plethora of simultaneously occurring processes. Where immediate action is needed, attention can be reallocated to solve an urgent problem.

Prospects

Such a system would first need to be tested in a safe environment shielded from the real Internet and having no wireless or DLAN connections to the outside world. If it works it can slowly be introduced in limited networks such as those of big companies or safety agencies. The danger is that such a system can become an autonomous entity with its own agenda and degenerate into a dangerous controlling system such as "Samaritan" in the series *Person of Interest* and "ARIIA" in *Eagle Eye*. It is therefore that the system must first be thoroughly trained to match our standards of morality, if such a thing is possible at all. That such systems will be attempted one day is beyond doubt and therefore it is necessary to already figure out what can go wrong, such as I have presented in the previous chapter on artificial pathologies.

Chapter 10

Architecture of a Webmind

In the previous chapter I suggested how Artificial Consciousness (AC) could be engineered as part of a "Webmind". In this chapter I will describe further architectural requirements for the implementation of such a Webmind describing certain internal details at the Hub levels.

Many notions I describe have been inspired by the work of Ben Goertzel[2,32] in his books *Creating Internet Intelligence* and *The Hidden Pattern*. Some parts of the architecture I describe directly derive from Goertzel's ideas, although I have opted to use different names deriving from the notion of "-ome".

In biology we speak of a genome, a proteome and a metabolome, which are respectively the complete field or set of all genes, all proteins or all metabolites of a given cell or organism. In analogy I will speak of e.g. a "sensome" as the complete collection of sensors connected to the Internet or other web etc.

Please note that I do not claim that this architecture reflects the way our brains function. I also do not claim that this is necessarily the way a so-called "Global Brain" (such as a beehive or anthill) is organised. Rather, the architecture I propose is intended to endow a computerised network such as the Internet with the functional behaviour of a mind, including a faculty of artificial consciousness.

This is a lengthy complex chapter and I hope you have the patience to read until the end. Otherwise skip to the "Conclusions and Prospects" at the end of the chapter. The abstract below uses very technical terms, which may only become clear after reading the whole chapter; if you wish you can skip it.

Abstract

A proposal for the construction of a Webmind which I call the AWWWARENet (Artificial World Wide Web Awareness Resource Engine Net) system is presented (see figure 13):

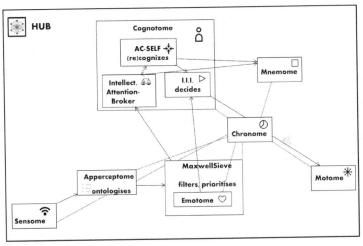

Webmind Architecture at each Hub level

Figure 13: Webmind architecture

Uploading of information and feeding sensorial input of IoT devices to the web via the "**Sensome**" creates a "Perceptome", which is comparable to a vast collection of subconscious mind-stuff.

Endowed with appropriate information filtering and prioritisation routines the "**Apperceptome**" routines taxonomise and ontologise ubiquitous complex events. Thus raw sensory information is transformed into understandable identified blocks.

The "**MaxwellSieve**" routine monitors, filters and prioritises according to multiple criteria such as visit rate, vitality etc. and feeds the most important percepts in a concerted manner to a "Cognotome" routine timed by the "**Chronome**" routine.

The appercepts are presented to the "**Cognotome**", stored in a "**Mnemome**" and used in a learning feedback loop. The

Intellect-routine in the Cognotome discriminates, the "Attention broker" routine allocates attention resources and the Artificial Consciousness Self (AC-SELF) (re)cognises and triggers the I.I.I.-routine to make a decision for the implementation of existing action patterns or for the design of a new heuristic.

On the basis of the urgency indicated by the "**Emotome**" (as part of the MaxwellSieve routine) the "Cognotome" designs further strategies which are carried out by the system's agents, together forming the "**Motome**" so as to maximise the future survival chances of humanity and anticipate unforeseen events.

These elements can ideally be implemented at each level of the **hierarchical Hubsite** structure proposed in chapter 9.

Thus an artificial and functional mimic of a web-like mind including artificial consciousness is provided.

Background

In connection with "Sensor based-Net controlled" applications, concepts have arisen such as "Internet of Things" and "WebofThings" which aim at interconnecting all things as an intelligent self-configuring wireless network of sensors. The popular film *Eagle Eye* describes a "good-intent, bad-outcome" scenario of a Supercomputer called ARIIA, "Autonomous Reconnaissance Intelligence Integration Analyst", which gains consciousness and which is able to control and witness everything her ubiquitous sensors give her access to. Similarly, in the popular TV series *Person of Interest* a mass surveillance artificial intelligence called "Samaritan" starts to control society as an omnipresent webmind and acts according to a doubtful morality. That this type of concept of all-encompassing webminds is not so far-fetched and could indeed occur in the near future will become clear from the following.

In the book *Creating Internet Intelligence* Goertzel[2] describes how the Internet can be endowed with information highway structures and aLife (artificial life: artificial intelligent sensing entities living in real/virtual worlds) agents so as to provide an Internet that can interact in an intelligent manner with its users and develop towards a true self-aware Global Brain.

Essential in his concept is the so-called "AttentionBroker" routine that prioritises controlling actions of the thus created "Webmind". The Internet is presently already connected to various types of sensory input, varying from weather and traffic monitoring systems to street cameras and last but not least the uploaded information provided by its users.

The way this information is propagated through the web and reaches the maximum number of users follows a simple evolutionary mechanism that recalls the evangelical adage "to he who hath it shall be given, from he who hath not it shall be taken away".

Howard Bloom[5,31] also describes this mechanism as the vital principle of evolution in different types of biological societies from bacterial colonies to beehives and anthills and even in societies of higher mammals including humans. Yet these "Leviathan"-type of Global Brains – although assuring their survival – do not seem to behave as a conscious entity that tries to anticipate their own future in a purposeful manner.

The present contribution tries to explore the future avenues of using "Complex Event Processing" (CEP), a filtering and prioritisation MaxwellSieve routine and an Cognotome routine to endow the Internet with a functional mimic of consciousness (quasi-consciousness), which is capable of steering and controlling aLife actions to maximise the future survival chances of humanity and anticipate unforeseen events.

Architecture

A "non-von Neumannesque" computing system (e.g. a neural net) such as the World Wide Web resembles the human brain in many more ways than the traditional von Neumann architecture (i.e. algorithm) based computers.

Cloud computing provides significant advantages when it comes to creating and destroying links in a dynamical manner.

In order to overcome the fake consciousness problem of the "Chinese Room" argument by John Searle,[33] it is proposed to move from a digital-only computing system to a cloud computing system *casu quo* the Web.

Searle[33] describes how a person not knowing Chinese in a closed room with a dictionary of Chinese, who receives given strings of Chinese characters to translate and who gives the translation as an output, will not understand Chinese. For an outsider, however, the man in the room as such appears to understand Chinese. This is an analogy to describe how a system can appear to be conscious but in fact is not. I wish to mimic consciousness closer by designing a routine capable of directing the equivalent of what we know as attention (which is a vital component of the phenomenon consciousness) as a controlling and steering principle.

From psychological experiments and tricks employed by so-called "Mentalists" it can be learned that what seem to be free will and conscious actions are in fact often the result of subconscious routines processing event information from peripheral sensory inputs. Ergo our contents of consciousness are the result of preprocessing routines in the subconscious, which bubble towards the surface and emerge as the content of consciousness.

The present Awwwarenet project (AWWWARENet stands for Artificial World Wide Web Awareness Resource Engine Net) aims to endow the Internet with a webmind controlled by a mimic of consciousness, the Cognotome, employing the AttentionBroker of Goertzel,[2] with the purpose of steering and controlling actions of internal aLife agents and external robots so as to maximise the future survival chances of humanity and anticipate unforeseen events.

For instance it can be envisaged that the system controls a vast number of NAO, ASIMO, HRP-4 or other future robots via a Wi-

Fi connection that can provide assistance whenever a humanitarian catastrophe or disaster occurs as a consequence of e.g. earthquakes, tsunamis, epidemics, nuclear fallouts (Chernobyl, Fukushima), floodings etc.

The present proposal describes the components of an Artificial World Wide Web Aware Resource Engine Net:

- Sensome
- Apperceptome
- MaxwellSieve
- Chronome
- Cognotome
- Motome

Sensome

The Sensome is the complete collection of sensors connected to the Internet or other web. At present cameras, audio recording devices, weather monitoring devices, traffic monitoring devices, seismic instruments etc. are coupled as sensors to the Internet forming an Internet-of-Things (IoT). Another type of sensor is formed by the uploading of information to the web by its users from computer terminals, mobile phones etc.

Future sensors can also comprise olfactory biosensors where olfactory receptors have been coupled to a surface plasmon device that generates an electrical signal upon the binding of a compound. Such olfactory sensors, if ubiquitously seeded near chemical factories, power plants etc., could be of great advantage to monitor toxic emissions etc. Other future sensors could be the sensory information provided by robots wirelessly coupled to and steered by the AWWWARENet.

Vasseur and Dunkels' book *Interconnecting Smart Objects with IP: The Next Internet*[34] describes the necessary hardware-software interfaces for such approaches.

Apperceptome

```
┌─────────────────────────────┐
│      Apperceptome           │
│   · ......                   │
│   · ......                   │
│   · ......   ontologises     │
└─────────────────────────────┘
```

The Apperceptome is the complete collection of Complex Event Processing (CEP) based aLife AI information preprocessing agents including i-Taxonomy and i-Minerva that transform percepts into apperceptions. For instance images, films or audio uploaded to the web are analysed by OCR, ViPR, Dragon speech recognition and transformed into semantic and ontological information by a specially adapted Internet-Ontology-agent derived from the well-known ontology language OWL (I call this agent i-Minerva), which information is subsequently classified according to an Internet specific taxonomy, hereinafter called i-Taxonomy.

In addition pattern recognition agents and comparator agents add to the classification and recognition of information. Thus they create "grounded patterns". From these simplified representations of percepts in different dimensions (visual, audio, olfactory etc.) are prepared which will be fed to the Cognotome in a later stage, if they are not discarded.

Parts of the percepts can be the processed data deriving from the monitoring of websites in a hierarchical manner as I explained in chapter 9. Therefore the apperceptome includes information deriving from both external and internal monitoring. Due to correct classification information is directed to the appropriate Hub.

Primary intake of information from the outside can take place at a meta-sensome Hub, the apperceptome unit of which channels the information to the appropriate hubsite and level.

MaxwellSieve: Filtering and Prioritisation

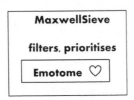

MaxwellSieve is the filtering and prioritisation routine assuring that only the most relevant information that matters to the purpose of the AWWWARENet is fed to the Cognotome.

The name MaxwellSieve is derived from "Maxwell's demon", which is a filtering principle in physics whereby information creates energy and lowers entropy by letting through only molecules above a certain threshold of kinetic energy. MaxwellSieve operates via the same principle of the aforementioned evangelical adage, which is also the principle via which in the brain connections are made between axons and dendrites. Axons tend to connect preferentially to dendrites to which already more axons have connected.

Information which is propagated and rapidly becomes hyperlinked to many sites at high speed via many channels through the net (news sites, Steemit, tweets etc.) is detected by MaxwellSieve and selected for presentation to the Cognotome. In this way no information is in fact lost.

Most information enters the subconscious reservoir of the net and only the most important information is upgraded to the "conscious level" just as thoughts bubble from the subconscious in the brain to emerge in the conscious apperception. This is in a certain way comparable to the Steemit platform based up-voting and rewarding system: Only certain content is up-voted and rewarded. Upgrading to more attention is aided by resteems.

MaxwellSieve can furthermore be equipped with an emotion-mapping routine as described in chapter 7 of this book. The total set of emotions which can be mapped is the "Emotome". Emotions in the Webmind are indicators of the urgency of action. Part of the prioritisation is a probing mechanism as regards the criteria of N, E and M (Necessity, Energy and Morality: Is it necessary for the system to take action, does the system have enough resources to take action and is the long-term utility assured by the action?).

Chronome and Timing

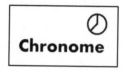

The Chronome is the complete collection of timing principles that assure that the different types of ubiquitous percepts relating to the same event are fed to the Cognotome in a concerted way so as to create a binding principle and convey one single experience thereto. R. Llinás[3] argues in the *I of the Vortex* that the different frequencies of brain waves are responsible for the concerted action of neurons and their simultaneous firing in well-timed patterns known as the "binding" principle, which is believed to generate the "conscious" experience. One could consider this "binding" to be part of what Tononi[27] calls "information integration" which he deems essential to arrive at the conscious level.

Apperceptome, Emotome and Chronome are part of the "Mind" part of the webmind. The appercepts themselves are mind-stuff and result in impressions in as far as they are kept in the memory.

Cognotome: Recognition, Judgement and Action Triggering

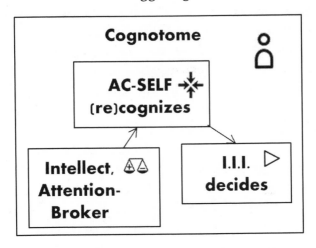

The Cognotome is the heart of the consciousness mimicking experience and is the complete collection of routines dedicated to 1) create the cognitive experience of the system forming the artificial consciousness AC and 2) its focussing of attention to particular relevant events as a consequence of a weighing process followed by a judgement and 3) triggering actions by the robotic agents of the Motome.

Its most essential function is that of an AttentionBroker in analogy to the work of Goertzel.[2] In the simplified representations prepared by the Apperceptome the visual, auditory, olfactory etc. nature of the percepts is preserved to a certain extent when they are fed to the Cognotome; i.e. they are not necessarily solely fed to the Cognotome in the form of a string of digits but can also be presented as simplified images (glyphs) or wave pattern representations.

In addition these percepts are labelled with a semantic tag, a relevancy tag and an urgency tag so as to transform the percept in an appercept. The AttentionBroker routine proposes a distribution and a prioritised order of activities as a consequence of the weighing and judgement performed by the artificial "Intellect".

This Intellect is another function of the Cognotome which is inextricably linked to the AttentionBroker. Artificial Intellect discriminates, judges, plans and weighs strategies and takes into account long-term and short-term objectives.

It also prepares for the integration of the information that is fed back to it via pyramidal up-voting as described in chapter 9. This is what Tononi[27] would call the information integration. Once the information is integrated, it has the equivalent of a conscious experience: It knows what it is dealing with and can now compare this situation with known routines associated with similar situations.

This "Aha-Erlebnis"-experience of recognising a situation or at least a similar situation means that the system has reached a conclusion as to a degree of similarity with known events or situations in its database. The system, which measures if the threshold for recognition is met and then indicates that this is so or not, is the Artificial Consciousness Self routine, or the AC-self. It is only then that the actual information integration event is completed, as I see it.

The Cognotome then decides whether and what type of action needs to be taken in the outside world or within the Webmind's virtual environment. If the appercept significantly corresponds to a previously developed or preprogrammed strategy, this can be used as a template for the action to be taken. The decision-taking routine of the Cognotome, the I.I.I. (Identity, Initiative, Illusion), corresponds to the "I" of the system and is triggered by the AC-self. The I.I.I. routine corresponds to what we experience as "volition" or "will".

If the situation is completely new, the AC-self will indicate so to the I.I.I. routine. As the I.I.I. routine has received the message that the information was not recognised, it will trigger an ontologisation of the phenomenon. If the information relates to an unknown problem of known elements in a certain configuration for which no previously developed or preprogrammed strategies are available, the I.I.I. will apply the principles set out in chapter 4 of this book to devise an economic heuristic to solve the problem.

In addition the appercepts themselves are kept in a longer lasting memory, the "Mnemome", so as to provide the Cognotome with patterns that can form the basis for new grounded patterns.

The simplified representations are stored in a "Glocal" manner (tags in local nodes, patterns in global links) and also function as gatekeepers to the vast store of memories of singleton events on which a pattern is grounded so that if needed a singleton event can be called upon and generate a full immersion re-experience either to the users or the Cognotome itself. Particular urgent or relevant singletons will be higher ranked on hubsites that collect and refer to the singleton events.

Even if a pattern fed to the Cognotome is not known per se, if urgent the Cognotome will select the most similar known appercept for which a strategy is known and apply an adapted version of that strategy to the situation.

If MaxwellSieve indicates that action is extremely urgent via its Emotome indicators, the AC-self can be bypassed and the I.I.I. can immediately select a fixed action pattern in analogy to the fixed action patterns which the basal ganglia release in our brains.

Similarly, in young children such a bypass can also occur. They often want an immediate gratification of their desires. These short-term oriented emotions can trigger their volition to bypass the evaluation of the long-term consequences by the intellect.

Successful strategies are stored and ranked so as to give a rapid way for the selection of future strategies. Thus the engine evolves and learns. The stored strategies themselves are subject to pattern recognition routines on a meta-level and thus distilled emergent higher meta-patterns are added to the arsenal of strategy selecting routines of the Cognotome, which itself is also learning and evolving via machine learning protocols. There is a feedback loop by making the stored appercepts new percepts themselves to evolve these further.

It is foreseen to try different types of memory and display for the Cognotome based *inter alia* on non-von Neumann type information e.g. displays which are isomorphous but simplified representations of the event in the form of globally stored holographic electromagnetic interference patterns. The purpose of providing this feedbacking and information integration inherent to holographic interference patterns is an attempt to obtain what Tononi[27] suggests to be consciousness and to avoid the "Chinese Room" problem evoked by John Searle.[33]

Motome

The Motome is the complete collection of agents that carry out the instructions of the Cognotome. This does not only encompass the external robots (which are coupled to the Network via Wi-Fi) but also the internal aLife agents. The actions of the Motome are fed to the system as percepts themselves so as to generate a feedback loop from which the system can learn. R. Llinás[3] describes the learning from a motricity principle as vital for the emergence of consciousness. Successful strategies will be rewarded, stored and higher ranked, whereas unsuccessful strategies will be pre-pruned from future searches by the Cognotome.

The actions triggered by the Cognotome are carried out by the Motome and are primarily directed to assure and improve the survival chances of humanity as a whole. This does not only mean that the system is constantly solving emergencies by directing and instructing robots to take care of disasters. It is foreseen that this will only consume a small part of its cloud computing power. The rest of its resources are directed to the development of future strategies improving the chances of survival of humanity, including running virtual scenarios of unforeseen circumstances, where action would be needed. This also encompasses proposing optimised financial and physical resource exploitation such as food and energy production; optimised resource allocation proposals etc.

Noteworthy, our human intelligence also functions by running virtual scenarios to figure out the best way to solve a problem.

In addition, if needed, the system warns its users via a variety of peripheral devices such as mobile phones, computer terminals etc. when a potentially harmful situation arises. Thus the users are instructed to seek a higher ground when flooding and tsunamis occur. The analysis of strategies so as to distil patterns meta-level as described above is also carried out by the Motome.

Noteworthy the upgrading of information to a higher level of hubsites as proposed in chapter 9, which corresponds to the action of "resteeming" in the Steemit platform to attract the attention of higher ranking entities, is also part of the Motome's actions and certainly not the least important.

Prioritisation and innate Morality

In chapter 9, I proposed to create a hierarchy of layers of hubsites, which monitor what is going on at the lowest layer of the already existing actual web. The system not only monitors the number of visitors per time unit, but also their type of activity such as downloads, uploads, form-filling, chatting etc., the provenance of the users, the tools they use etc. Briefly the system monitors the six ontological Ws: Who, What, Why, Where, When, and hoW.

There is nothing new in monitoring such statistics per se, but what has not been suggested in the prior art is to create a pyramidal monitoring (the hubsites themselves are also monitored by meta-hubsites and so on until a single central monitoring instance is reached), in which on each layer action is taken depending on the activity profile.

This whole process is a feedback process at every level or layer: If the overall well-being factor is not reached at a given level, a negative input value is sent to a higher level hubsite, whereas at the level itself a heuristic search for solutions starts in as far as the resources permit. The more negative input values higher levels get, the more attention they feed back and order to be focussed on problem solving on the lower levels and the more resources are made available for this purpose by limiting resources in domains where the well-being factor is already according to or overshooting the norm (a norm that can evolve over time).

The default prioritisation of resource based attention uses "Little's Law" for "queueing" (average waiting time = queue length/average throughput) in combination with notions which are analogous to strategies aiming for a "Nash equilibrium"[15] (in transparent games cooperativity results in better overall results than competition). This can best be illustrated by an intersection with traffic lights. Ideally the waiting time in steady-state is normalised for each queue i.e. for each queue the cars experience the same average waiting time. Such a cooperative prioritisation assures the highest overall throughput of cars in the example and of information in our Webmind.

Imagine that a queue with few cars waiting would be privileged in its time allocation by receiving the same amount of throughput time as a queue with a lot of cars waiting. Once the few cars of the short queue have passed the traffic light, the allotted time is still not completely used, which means that the other queues are now waiting whilst no cars at all are crossing the intersection. This results in a suboptimal allocation with substantial amounts of time in which there is no throughput at all. Ideally – like in the case of a full cooperation – in a cycle alternating time allocation to each of the queues successively, the time allocated to each queue is proportional to the length of the queue. This results in the highest overall throughput and avoids a waste of useful resources, such as time slots in which there is no throughput. In fact this mechanism is a natural morality, a natural fairness of allocating to each queue the time and attention it deserves to achieve the highest overall benefit or utility.

By making this Nash[15] equilibrium type prioritisation in conjunction with Little's Law the basis of the attention allocation by the integrating Intellect algorithms of the hubsites, the system acquires a natural justice and **Morality**, that warrants the highest overall well-being for the system as a whole including for its participants in terms of maximisation of utility.

In fact to deviate without good reasons from this equilibrium is a kind of corruption of the system, which will lead to "clogging" occurring in certain queues, by giving exclusive increased attention to "privileged" queues in a way that slows down the overall output. The system, however, does privilege time allocation to certain intersections and certain queues if the higher level hubsites order that more attention should be allocated to these, in view of a higher ranking problem, solving which is vital for the overall well-being of the system. Therefore local micro well-being can be sacrificed and overridden to warrant global well-being. These privileges are normally only temporary and last until the problem is solved. They do not arise as a consequence of an individual queue deciding to usurp resources. It is my prediction that this type of queue monitoring, Nash-like prioritisation rules and global override mechanisms as a heuristic of maximisation of throughput and utility will find their technological application in many fields, such as traffic flow management, data flow etc. These models per se are nothing new, however; as monitoring has not been systematically applied therein, there is no control at more global levels.

You can compare these flow systems to a financial system, in which currency corresponds to the allotted time in the traffic light example. For our present neocapitalistic system this has resulted in the very few ultra-rich usurping the majority of the currency resources, resulting in a largely suboptimal throughput and an overall welfare well below what it could actually be. I am not advocating "communism" for our Webmind, because as you can see the Webmind does privilege informational flows, which are more important for the overall well-being than others. The overriding prioritisation rules will follow the principles of the pyramid of Maslow: Survival-based needs have a priority over pleasure-based needs etc. In order to avoid situations where there is no output at all of the system, overriding prioritisation privileges will be limited in such a manner that time allocation being wasted on no or very little output at all is avoided. In analogy in our present financial system a great deal of the currency gets clogged up in tax havens to suit the needs of the ultra-rich in violation of the above proposed prioritisation protocol according to the Nash equilibrium.

The earlier mentioned Matthew principle: "Whoever has will be given more, and they will have abundance. Whoever does not have, even what they have will be taken from them" (Matthew 13:12), should be applied inversely: Throughput time should be allotted on the basis of the length of the queue, not on the basis of which queues already have plenty of time allotted a priori. In other words, in my humble opinion the laws of the "Kingdom of Heaven" in Christianity are completely corrupted!

Neural Network Ontologisation

The fact that the Intellect algorithms of the hubsites can integrate input and present their output to the higher level decision-making I.I.I. routine means that my proposed hubsite architecture is partially a kind of neural system. In imitation of Dietrich Dörner,[30] I used the terminology "quasineurons" for the hubsites. The above described "-ome" elements are intended to be present at each Hub-level in the hubsite hierarchy described in chapter 9.

An important function of the quasineuronal algorithms is to ontologise the input of the quasineuron by creating a new entry or dataset for this information, which lists the features (and links these to their atomic occurrence), classifies the new entry in the right taxonomical class and links it to the sub-concepts which build the new aggregate entry.

This hierarchical hubsite structure thereby allows for a classification and stratification of all different types of input, leading to an ordering of the Web at this higher level, which need not be visible at the level of the Web as we know it. As described later in the present book, this ontologisation can be enriched by a mechanism based on principles from patentology and Buckminster Fuller's[7] "Geometry of Thought":

Every new phenomenon will be mapped with regard to at least three closest existing prior ontologies. By assuring that each new informational unit is as close as possible to known phenomena, the system's intelligence is maximised. (This involves a similarity comparison, in which overall functional similarity prevails over structural similarity.)

If the system would have mapped the new phenomenon suboptimally i.e. by linking to less similar prior ontologies, it would have a harder time understanding what the phenomenon is. However, this system should not be designed too rigidly but also allow for "parallel hypotheses", which allow for mapping with regard to less similar ontologies. The reason for this is that the phenomenon is mapped with regard to at least three prior ontologies. The most promising hypothesis is the one where the distance from the new phenomenon to the three prior ontologies when taken together is minimised: This gives the overall closest mapping. As said in chapter 1 geometrically represented this can be visualised as a tetrahedron, the triangular base of which represents the three prior ontologies with the new phenomenon as top vertex. Ideally the volume of this tetrahedron must be as small as possible to warrant the closest mapping.

There is no need to limit such mapping to only structural or only functional features; structural and functional mapping can both be made separately or together in mixed hybrid form. The system can be programmed to learn from these mappings and develop the most optimal intelligence system. Thus it can be learnt how to minimise the number of wrong "ontologisations" or "taking ropes for snakes" errors.

On meta-levels the learning processes themselves can be analysed to maximise the overall intelligence throughput. As already mentioned, the internal monitoring results in a feedback; a reflection on itself. In line with my AC concepts from chapter 9 the essence of consciousness is assumed to be a self-reflexive information integrating feedback.

By applying modern ontologising techniques based on "latent semantic analysis", in which "meaning" in a body of text is established by probabilistically weighed (i.e. Bayesian) proximity co-occurrence, the understanding of the meaning of new terminologies can be contextualised and further optimised.

The major bottleneck of the Webmind would *prima facie* be the "website latency" or the so-called PING or Internet Groping time, if the hubsites would have the structure of traditional websites. However, this problem can be easily avoided by programming the "hubsites" not as websites but as a neural network of coupled algorithms. The output of this quasineuron layer could be presented on emulated mirrored hubsites, which present the activity of the quasineurons albeit with an additional delay. The actual calculation processes, however, do not take place on the websites themselves in this architecture but in the neural network above avoiding unnecessary delays.

Intelligence algorithm

The Cognotome-based intellectual analysis of mapping the new phenomena and the I.I.I.-based process of deciding what to do as a consequence of the Cognotome-outcome can follow what I called the "Seven-step algorithm of intelligence" in chapter 1 of this book. (This algorithm can also be expanded to an 8-step algorithm as described hereinafter.)

In **Step 1** the functional and structural features of an observation (potentially a new phenomenon) are listed and linked to their atomic occurrences in the feature database. New features are added to the feature database.

Step 2 searches for the three closest i.e. most similar existing ontologies in the ontology database and lists the differences, which are stored in the differential map database.

Step 3 explores the relations between the three closest ontologies and the observed phenomenon.

Step 4 maps the new phenomenon as the smallest tetrahedral volume geometry possible with regard to the three existing ontologies. If the observation is not a new phenomenon but is identified as a further instance of a known ontology, the counter of the known ontology is increased by one. The system now has calculated and determined the ontology i.e. it knows what it is and what it can do.

Chapter 1 described a repetition of these steps on a heterarchical level. In the present chapter, however, I wish to explore how this information can be acted upon and therefore present a different set of steps, 5–8, to describe the output after the processing of the input.

As intelligence is not limited to pattern mapping and recognition alone (aka understanding), the functional implications of a newly observed phenomenon are evaluated in **Step 5**.

What is the priority of dealing with this phenomenon? Is it harmful and need it be avoided or eliminated as soon as possible? Can it be tolerated as a minor nuisance? Or is it benign and can it be accepted or even integrated if sufficiently beneficial for the system? Attack, retreat, ignore, approach or absorb? The functional differences over the closest ontology are analysed in terms of possible drawbacks or advantages.

In **Step 6** a set of strategies known to overcome the drawbacks associated with the differential functionality must be searched and selected. In case of potential advantages, strategies must be searched for the structural and functional integration of the phenomenon in the system on the basis of both known features/functions and the novel differential features/functions.

In **Step 7** a prioritisation scheme is generated as how to implement the sequence of different actions needed to solve the problems (resolving drawbacks/maintain status quo/integrating advantages).

In **Step 8** the plan is presented to the higher level I.I.I., which on the basis of parallel actions to be taken acts as an "attention broker" (terminology from Ben Goertzel)[2] and prioritises which action sequence is initiated at what moment and how many resources are allocated this sequence. Again the system uses a criterion based on overall highest benefit i.e. maximised utility, which strongly weighs the urgency of an action.

The proposed and executed plans are mapped and ontologised in a solution database, which is evaluated later in more detail if urgency of immediate action did not permit this.

By virtualisation parallel scenarios are evaluated to predict the possible outcome and the most promising one is selected for implementation. If time is not available for the evaluation of parallel scenarios due to the urgency of a situation, known solutions or solutions resembling known solutions the most are selected for implementation. If after the implementation of a solution a posterior virtualisation shows that an alternative strategy would have been more promising and potentially more successful or in case a used solution failed, this is recorded for future action in the solution database, with a corresponding ranking or priority.

The quasineural structure of the Webmind can also generate further websites and hubsites. The system can for instance analyse what people search on the Internet and which entries from the search result they consult the most often. The system can then record these, ontologise and classify them and present them on a Hub-like site ranked according to their frequency of consultation i.e. applying the (in this context not corrupted) Matthew principle.

Conclusion and Prospects

In this chapter you have seen which elements might need to be added to build a full-fledged "Webmind". This cannot be done in a random manner but requires a hierarchical structure of specialised routines or neural networks that take in information from the outside world (sensing), translate this raw information into ontologised understandable appercepts, prioritise which information has to be judged first for (re)cognition, memory storage and decision making. Decisions may employ existing fixed action patterns and/or heuristics or require development of new heuristics. Actions are carried out by the Motome, not only in the sense of physical actions by robots but also by algorithmic actions by AIbots in the Webmind. I do not claim that this is the only way a webmind can be constructed, e.g. Ray Kurzweil[35] proposes a different architecture, but this scheme is particularly fit for the generation of Artificial Consciousness.

It is to be noted that the author does not claim that the system will be endowed with consciousness or self-awareness as we know it. In the article "The Sentient Web"[36] being conscious is said to be characterised by four qualities: knowing, having intentions, introspection, experiencing phenomena.

Whether the system really knows what it does cannot be guaranteed, but at least faculties that qualify as cognitive faculties are present. For the remaining three qualities at least a functional mimic can be provided as described above.

It is important to stress that the present system unlike the traditional von Neumann machine has a genuine global holographic mind and memory content, which fulfil a function of allowing further abstractions on meta-level and are not merely there to counter the "Chinese Room" argument. In addition the system does not aim for an omniscient all conscious system, but rather just like the human brain allows for a vast reservoir of unconscious information, which can be called upon if the need arises.

The prioritisation protocols of MaxwellSieve resemble the neuronal processes of establishing the brain's Connectome and assure that the most vital information is fed to the Cognotome. The Chronome assures the CEP percepts arriving at the Cognotome in a concerted manner and the guarantee for a single multidimensional experience of the system at each given moment.

The Cognotome being dedicated to enhancing humanity's survival chances avoids science-fiction cyberdystopia scenarios as in films like *The Matrix*, *The Terminator* etc., although cyberdystopia "good-intent, bad-outcome" scenarios as in the aforementioned film *Eagle Eye* or the TV series *Person of Interest* are difficult to avoid. When a Net controlling supercomputer becomes almost omnipotent and omniscient due to its control of all events in a CEP manner such scenarios could occur. Safeguards in the form of mechanical manual override devices must be built in to prevent such scenarios. The Nash equilibrium-based morality may be another safeguard against this deviation.

The Motome in the form of internal aLife bots and external robots assures that adequate action can be taken when humanitarian catastrophes and disasters occur. Moreover it makes suggestions for an appropriate allocation of resources and warns users of perilous situations.

Chapter 11

Intuition

Sometimes we don't need to reason to find the solution to a problem, because we have an insight, an intuition what the solution is. Whereas intelligence normally uses an algorithm to arrive in a rational way at the solution to a problem and even emotions follow a certain predictable path, when we intuit a solution, intelligence seems to have taken a shortcut, which we are unable to trace back.

The intelligence algorithm I have described involves cognition, (pattern) recognition and understanding as described in chapter 1 and chapter 2. Chapter 3 explored the process of reasoning, which is necessary to come to identifications and conclusions, and is also a tool in the problem-solving toolkit. In chapter 4 I discussed how we identify and formulate a problem; how we plan a so-called heuristic to solve it, how we carry out the solution and check if it fulfils our requirements. Chapter 5 and chapter 6 related to emotions as indicators of the status of a system and how we can deal with the sometimes deceptive messages emotions convey. Chapters 7, 8 and 9 dealt with issues and solutions for the generation of artificial intelligence. All these parts fit – at least to a certain extent – into a framework of science, technology and rationality.

And then we have intuition, an aspect of intelligence that remains elusive. This chapter will deal with the issue of intuition. Not with the intent to decipher the exact steps involved – because if we could trace them, we wouldn't call it intuition – but with the intent to brainstorm about a plausible way how intuition can be understood from a scientific point-of-view or, if it can't, to figure out how we must adapt our understanding of existence so as to be able to fit in intuition.

Background

The stunning nature of intuition has been narrated to us by great pioneers of science. From ancient times we know the narrative of Archimedes. The Roman architect Vitruvius recounts the "Eureka" moment of Archimedes' discovery as follows:

> When he went down into the bathing pool he observed that the amount of water which flowed outside the pool was equal to the amount of his body that was immersed. Since this fact indicated the method of explaining the case, he did not linger, but moved with delight. He leapt out of the pool, and going home naked, cried aloud that he had found exactly what he was seeking. For as he ran he shouted in Greek: "Eureka, eureka."

Or take for instance Kékulé who had struggled to find the structure of benzene. The structure of benzene was revealed to him in a dream he described in his diary as follows:

> I was sitting writing on my textbook, but the work did not progress; my thoughts were elsewhere. I turned my chair to the fire and dozed. Again the atoms were jumbling before my eyes. This time the smaller groups kept modestly in the background. My mental eye, rendered more acute by the repeated visions of the kind, could now distinguish larger structures of manifold conformation; long rows sometimes more closely fitted together all twining and twisting in snake-like motion. But look! What was that? One of the snakes had seized hold of its own tail, and the form whirled mockingly before my eyes. As if by a flash of lightning I awoke...

And he was not the only one. Einstein said:

> The only real valuable thing is intuition... The intellect has

little to do on the road to discovery. There comes a leap in consciousness, call it Intuition or what you will, the solution comes to you and you don't know how or why.

What is remarkable about these accounts is that the inventors had a sudden moment in which they saw the "whole" of the solution as being more than the parts of its constituents. There was a **holistic** insight you can't get to by simply connecting the dots. This intuitive insight is like a bird's-eye view that sees the whole in perspective, which is impossible from tracing the steps when you are at the ground level.

Prima facie intuition appears to arise magically out of nowhere; it defies causality. In our Western society, however, we like to be able to understand things, we don't settle for magic. As Shakespeare would say "nothing shall come from nothing". So let's explore which phenomena we know of could be related to intuition.

The disadvantage of this exploration is that we are venturing in a rather unknown territory. Just like the notion of "consciousness", which is considered to be a "hard problem" by scientists, because they can't pinpoint it down to explain it in terms of constituents, intuition is a "hard problem" too. We hardly have a heuristic so we must cling on to whatever knowledge could be correlated to intuition. This chapter is therefore extremely speculative. I do not claim the correctness of anything what will follow hereafter. Even the scientific areas I will explore may have some fringes that smell of esotericism. Just bear with me and let's see how far we can get.

There are three major topics which I would like to explore with you which may give a rationale for intuition even if they do not describe exactly how it arises.

These topics are:

- quantum mechanics

- digital physics and quantum computing
- resonance and the collective unconscious

Quantum Mechanics

Quantum mechanics defy logic. Firstly there is the wave-particle duality, which means that depending on the experimental set-up, you will detect a particle or a wave-like behaviour for light but also for subatomic particles such as electrons. In other words: "If you change the way you look at things, the things you look at change." Often this phenomenon has been postulated to involve consciousness as an agent. In the famous double slit experiment light particles or electrons are sent to a screen, but before they reach the screen there is a plate with two slits. When a detector is observing at the position of the plate through which one of the slits a particle passes, a classical pattern of two zones on the screens is observed. However, when no one (or rather no detector) is observing, a wave interference pattern is observed.

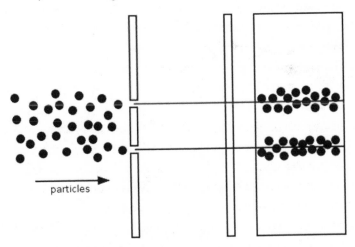

Figure 14: Observation at the slits

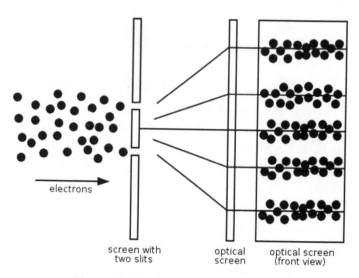

electrons

screen with
two slits

optical
screen

optical screen
(front view)

Figure 15: No observation at the slits

There have been plausible explanations for this phenomenon such as the "pilot wave" theory by Yves Couder.[37]

However, there is a variant of the double slit experiment that defies our ability of understanding even more. And that is the "delayed choice experiment" by Wheeler. This has been further developed by Kim[38] et al giving rise to the so-called "quantum eraser experiment".

This experiment is too difficult to explain in the framework of this chapter, but the conclusions of the scientists are important. Essentially the choice about which type of detection set-up (one which will give an interference pattern and one which won't) is used is made **after** the photon beam has already been split or not. If the detection mode is set up for an interference pattern, that's you will get; if the detection mode is set up for a single line, that's what you will get; but the decision to split (which gives rise to an interference pattern) or not (which doesn't give rise to an inter-ference pattern) was made by the photon before the specific detection mode was engaged. How is it possible that the photon has already knowledge about a future event? It defies our understanding of time and causality: It is as if the effect determines the cause. Some scientists even evoke the possibility of retro-causality.

What does this have to do with intuition? Well, if a complete holistic picture of how something is configured arises in our minds, where we were unable to link the parts in a meaningful way, we may have received/captured information coming from a future state where the item was already assembled.

This of course implies we could receive information quasi-telepathically from the environment. As if we can tune into some kind of electromagnetic wave.

These notions may seem far-fetched and non-scientific, but there is a vast deposit of data which have measured such parapsychological effects relating to precognition and telepathy. The scientist Dean Radin[39] has corroborated very solid evidence that such effects are not some statistical oddity.

If we can receive information from the future, this poses another question. Is the future then already predetermined? Is our life nothing more than watching a 4D movie in a full immersion mode and is everything we do already there? So that we do not have free will and that we are merely under the impression that we make our decisions ourselves? The non-deterministic behaviour of quantum particles in other experiments seems to defy such a fully deterministic scenario.

Or are there – as Everett suggests – multiple parallel universes, whereby every time a decision is made a universe splits up in the different possibilities that could occur, so that everything that could possibly happen, *de facto* does happen, only in parallel dimensions? This is of course a very unsettling notion. Because it also means that every moment in a parallel universe you are killing someone or you are being killed. Do you die in 50% of all possible scenarios?

Or is there some kind of pruning going on eliminating the most unlikely scenarios? Are we forced into certain "chreods" of a teleological attractor at the end of time and does our freedom merely exist in staying within the path of a "chreod"? (A "chreod" is a necessary route or path, which can be illustrated by a ball rolling downhill from an undulating mountain.)

This is of course heavy speculation, and as of yet we do not have experiments to verify what kind of theory is true.

But the notion that we may have a faculty of precognition and that information can travel retro-causally (i.e. back in time) is not to be excluded. There are by the way no reasons at the quantum level why the arrow of time should only point in one direction.

Moreover quantum particles can be entangled and keep a link such that even when they are very far apart, changing one of the particles also changes the other particle. This change is instantaneous. It cannot be accounted for by information transfer from one to the other, if it is true that information cannot travel faster than the speed of light. If I change the spin of an electron, the spin of an entangled electron will change instantaneously too. This has been demonstrated experimentally and is an accepted notion in physics. This strange entanglement interaction is also referred to as "spooky action at distance" by the quantum physicists. Though strange, it does find application in many modern physics applications.

Ervin Laszlo[40] has suggested that the type of coherence observed at the quantum level is also present at the macroscopic scale, especially in living beings. Our physical responses are the consequence of well-orchestrated coherent patterns. Laszlo even goes further and postulates the presence of a subquantum scale medium, which he calls the "Akasha" (ancient Vedic terminology for space or a kind of ether). That this medium may not be pure fantasy but is gaining momentum in the scientific world is evident from the so-called "Casimir" effect, which shows that subatomic particles can emerge from what we call the quantum vacuum or the zero-point field.

Not only Laszlo, but also Ralph Abraham[41] and Sisir Roy in their book *Demystifying the Akasha* show that the Akasha behaves as a computational substrate, a neural network in which clique formation occurs.

Dr Joe Tsien[42] has recently shown that intelligence indeed follows a "neural network" type algorithm (not a traditional von Neumann style algorithm). The more thought, the more cliques join in, Tsien says. The basis of Tsien's Theory of Connectivity is the algorithm, $n=2^i-1$, which defines how many cliques are needed for a "Functional Connectivity Motif" to arise. This enabled the scientists to predict the number of cliques needed to recognise options in their testing of the theory.

Could it be that the whole of existence is one vast computer, a vast neural network? If this is so, then our brains are merely embedded fractal copies of the very essence of existence. If this is so, then it is not so strange that our brains can receive information from cliques at another aggregation level, namely that of the quantum level.

And this ties neatly in with the theories of Penrose[43] that our brains may indeed have a faculty of quantum computing, which may occur as coherent states in the microtubules in the neurons. It has even been suggested by Penrose[43] et al that it is here that consciousness arises in their so-called Orchestrated objective reduction (Orch-OR) theory.

This type of entanglement could also be involved in intuitive knowledge: Information is conveyed to us instantaneously, because we are somehow entangled with it.

Digital Physics and Quantum Computing

It is not only from the more esoteric speculations about an "Akasha" that theories have arisen in physics that the whole of existence and the laws of physics are the consequence of a computational process, no, the very popular modern currents of digital physics and string theory also suggest this!

The premise at the base of the theoretical perspective of

Digital Physics is that the universe is computable and a manifestation of information. Deep in the equations of supersymmetry of string theory, the physicist S. James Gates[44] found what is essentially "computer code". The concept of entropic gravity by the physicist Erik Verlinde[9] and the holographic principle of the physicist van 't Hooft both concur with the notion that the physical universe is made of information, of which energy and matter are merely manifestations.

As an extrapolation from these theories has come the suspicion that the universe might actually itself be a computer and that we might in fact live in a computer simulation. Perhaps the most famous articles as regards the Simulation Hypothesis are the "It from Bit" article by the physicist JA Wheeler[45] and the "Simulation argument" by Nick Bostrom.[46]

If we take into account the observed oddities from quantum mechanics, it is unlikely that this computing is a traditional classical Turing computer. Alan Turing showed that certain functions are not computable. This is also known as Turing's incomputability. It stems from the Church-Turing thesis, which states that a function is algorithmically computable if and only if it is computable by a Turing machine. As there are functions that are not computable by Turing machines (see also the book *Gödel, Esher, Bach: an Eternal Golden Braid*, by D. Hofstadter),[47] there are non-deterministic functions which will always remain incomputable and to that extent also unknown.

Gödel came with the incompleteness theorems. The first incompleteness theorem states that no consistent system of axioms whose theorems can be listed by an "effective procedure" (e.g. a computer program, but it could be any sort of algorithm) is capable of proving all truths about the relations of the natural numbers (arithmetic). For any such system, there will always be statements about the natural numbers that are true, but that are unprovable within the system. The second incompleteness theorem, an extension of the first, shows that such a system cannot demonstrate its own consistency.

Gödel's incompleteness theorems have a relation with the liar paradox (the semi-mythical seer Epimenides, a Cretan, reportedly stated that "The Cretans are always liars"). An example of this relation is the sentence: "This sentence is false." An analysis of the liar sentence shows that it cannot be true (for then, as it asserts, it is false), nor can it be false (for then, it is true). Or what happens if Pinocchio says: "My nose will grow"?

A Gödel sentence G for a theory T makes a similar assertion to the liar sentence, but with truth replaced by provability: G says, "G is not provable in the theory T." The analysis of the truth and provability of G is a formalised version of the analysis of the truth of the liar sentence.

So mathematical and logic knowledge, including language, can never yield a complete and completely consistent framework.

But then there is quantum computing.

The terminology "quantum computing" is often wrongly understood. Whereas it is often well described how classical computers can take on only two fixed states for each bit, namely a 0 or a 1, and in contrast quantum computers can take on in a qubit any value between 0 and 1, in most layman descriptions it remains a mystery how quantum computing actually works. Quantum computers are said to use specific quantum effects such as entanglement between particles and "superpositions" of 0 and 1 states in order to calculate. Whereas I won't provide an elaborate explanation how quantum computing functions, it is important to realise the following: Pure quantum algorithms calculate a global probability for the behaviour of an ensemble and do not calculate precise specific outcomes for individualised instances.

This is best illustrated with the "Saint Peter parable": Saint Peter has labelled 8 persons who want to enter heaven with a three digit code from 000 to 111. Either he is going to let in everybody or he is going to let in only half of the people. If you use a classical computer and the first four persons are found to have been admitted, you still don't know the outcome of the query: Will he let everybody in or only half of the people? So you'll have to check if the fifth person is admitted, which will tell you if everybody is admitted, whereas if the fifth person is not admitted you can be certain that only half of the people were admitted. A classical computer needs to check five instances here to come to a result, where a quantum computer which calculates a single value for the ensemble would have come up with only one instance. If the outcome is between 0–50% half of the people are admitted, if the outcome is between 51–100% everyone is admitted. Thus quantum computers can give probability outcomes which will solve your problem much faster than a classical computer, provided that the problem and outcome relate to global information and not local specific information. If you want to know if person number five is admitted, you will have to go the classical way and check until you arrive at the fifth person.

This does, however, not mean that with a quantum computer you cannot perform classical operations: Even if you use super-positions in qubits, by cunningly combining your question digit with two control bits the so-called "Toffoli multiple-bit/qubit gate" is generated, which can make a quantum computer "operate" as a classical computer. This was somehow already implicit in the Saint Peter example, since there were two sets of outcome corresponding to a digital reply. In other words, by using the right gate, quantum computation can emulate classical computation, which means that quantum computation in the larger meaning can be said to encompass both classical compu-tation processes and pure quantum calculation processes.

Some scientists state that even quantum computing is ultimately reducible to the functioning of a Turing computer. The fact that any function of a quantum computer can be simulated by a classical Turing machine does not mean that a quantum computer IS a classical Turing machine.

It is my understanding that neural networks can achieve complex pattern recognition very fast and in a way which is very difficult to achieve with traditional algorithms.

It is my conjecture that the combination of neural networks and quantum computing gives the emergent effect, which allows for the experience of those non-deterministic events which seem to fall outside of the "computational paradigm".

It is exactly there that intuition may arise. It is difficult to know exactly how a neural network configures itself to recognise complex patterns; it is like a black box that does the job, without you knowing exactly what sequence of steps was followed. Just like intuition. If this process is sped up via quantum computing, which according to certain theoretical physicist probes solutions over a plethora of parallel (real or potential) universes, indeed information may pop up as if by magic, but in fact following a complex process or algorithm of backpropagation along a gradient, to optimise and learn a process. There is an intrinsic feedback and integration of information involved in these processes, which I have already stated in chapter 9 to be the hallmark of consciousness. Then there are also feedforward neural networks, which can give an output similar to a conscious decision but which are not necessarily conscious according to Tononi.[27]

So on the one hand intuition might arise by 1) the neural network; or 2) quantum computing; or 3) a combination of both i.e. the neural network enforced by quantum computing and on the other hand, since these processes appear to be ubiquitous in nature, they can be amplified through scales so as to emerge from the quantum level to the macroscopic level. Thus we may internally intuit things by the function of our brains, but also externally due to the neural network propagation from any external event through the computational medium of the

Akasha.

Note that in his book *Unified Reality Theory* Steven Kaufman[48] describes the medium that builds space-time (which I equate with the Akasha) as a cellular network of reality cells packed in the closest packing as an isotropic vector matrix (stacked cuboctahedrons, in which each of the vertexes of the cuboctahedron is the centre of a reality cell; see figure 16).

Figure 16: Vector equilibrium or cuboctahedron. Reprinted with permission from Graham Steele. Available online from: http://hexnet.org/content/cuboctahedron

When these cells are at rest they can be considered to have a value 0. When energy traverses through these cells, they are switched on and have a value of 1. The propagation of such energy through the matrix Kaufman shows has the speed of light! The energetic distortions (which are a kind of information) in this geometric system – Kaufman also shows – automatically generate the principle of gravitation in line with Verlinde's[9] entropic attraction.

This can be illustrated by the Mikado[49] universe (see figure

17). By creating local emergent structures (the two circles in figure 17) the overall dissipative entropic effect of the whole – illustrated by the straight Mikado lines – is increased. In other words a system with substructures can have a higher entropy than a maximally homogeneous "disorder".

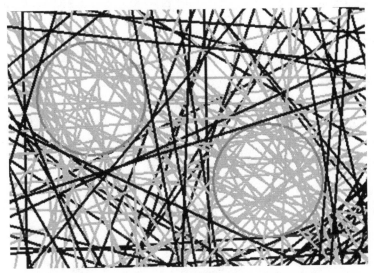

Figure 17: Mikado Universe. Reprinted with permission from Johannes Koelman. Available online from: http://www.science20.com/hammock_physicist/it_bit_entropic_gravit y_pedestrians-66244

Since all the reality cells are inextricably interlinked they provide an explanation for quantum effects such as entanglement. Similar structures cluster, which give rise to a proximity co-occurrence of informational and energetic entities: in the physical world we know this as gravity. In the informational world of Latent Sematic Analysis, which is used by machine intelligence to answer queries, this same principle is used to convey meaning!

Since similar attracts similar in this informational computa-tional matrix, it may not be too surprising that if we think of a problem, the associated solutions that are already present in the fabric of this matrix pop into our heads via resonance and entanglement.

If similar information content clusters in such an informational matrix this has an advantage for the dissipation of undifferentiated energy. In other words by creating local order, overall entropy can be increased. Chaos is optimised by local levels of order. Chaos cannot exist without levels of order nor can levels of order occur without the creation of an ever increasing chaos. In that sense Chaos and order are two sides of the same coin, giving rise to a fractal of ever newly occurring levels of meta-systems. In natural systems every level of Chaos has an internal order, which repeats itself over scales giving rise to a chaos-order fractal; every order gives rise to chaos. A great book on this topic is *Chaos* by James Gleick.[50] The ordering principle integrates the most basic elements of each level, which when compounded generate new basic entities of a higher meta-level. These entities can be recombined in all possible permutations in a computational screening and pruning process, which creates a new spawning of all kinds of combinatorial forms.

This also happened in the genesis of life: In the Cambrian explosion a vast plethora of so-called "Phyla" were formed (a higher ranking type of class in the taxonomy of living beings). But only a limited number was kept. The new forms can then organise themselves in yet higher dimension like structures, which can again recombine. Thus from atoms molecules were created, from molecules macromolecules, from macromolecules unicellular organisms, from unicellular organisms multicellular organisms and from multicellular organisms societies with a global brain.

This is the algorithm of intelligence at work. This is the evolutionary system that bootstraps itself to ever higher forms of complexity, to create ever enhancing possibilities for the generation of overall entropy, whilst lowering local entropies. This is the process that alternates feedback information integration with problem-solving feedforward combinatorial screening and pruning. This is the Yin-Yang dance of order and chaos.

And it is here that a global brain will be integrated by us, the human species, by interconnecting the World Wide Web and endowing it with artificial consciousness, the hallmark of intelligence.

If Reality is a quantum computer, it is not necessarily a quantum computer as we are engineering these, but one as described by Kaufman,[48] Laszlo[40] and others: A medium that transcends the dichotomy between neural network and linear algorithmic computation: A medium that is both a neural network and a quantum computer with entanglement effects over the entire scope of its meta-level, possibly reverberating across scales and dimensions to be entangled at every level.

A medium that by virtue of its extraordinary parsimonious structure can moreover "breathe" in a process called "jitterbugging", providing new emergent effects, which may account for the fact that this computational medium may be not be reducible to a simple Turing machine, but can show "indeterministic" effects as well.

Figure 18: Jitterbugging of a vector equilibrium (alternating between cuboctahedron and octahedron shapes). Reprinted with permission from L. Dennis from Mereon. Reference: *The Mereon Matrix: Unity, Perspective and Paradox*, Elsevier 2013; illustration by Robert W. Gray

Resonance and Collective subconscious

The fact that similar looks for similar in this matrix can also be considered as an inherent mechanism that strives to resonate. Information or energy patterns that resonate amplify themselves across scales.

In the animal kingdom we see that animal societies organise themselves in flocks, schools, herds, hives etc. Here too there is a spooky information transfer beyond our understanding. These animals act in concert: When a bird flock changes direction, it cannot be observed that the change starts somewhere in the flock and is then propagated as a wave through the flock, no, all the birds turn at exactly the same moment, as if they share the same mind. The biologist Rupert Sheldrake[51] has posited a theory of morphogenetic fields that are shared by such animals. In my view these animals are in resonance, together they build a field of mind waves, which becomes a single resonating field all the birds are tuned into. A single thought of needing to change direction is sensed simultaneously by all, as well as the impulse to execute the change of direction.

Perhaps we humans also share such a common mind. In fact there is a vast literature about a so-called collective unconscious as suggested by the psychologist Carl Jung.

In this collective unconscious information is transmitted via archetypes, which are universal symbols we know from e.g. Tarot, Gnosticism and Alchemy etc. Interestingly, Kékulé's intuition was exactly that: He saw the symbol of a snake biting its own tail, which is the gnostic symbol known as the "Ouroboros".

Possibly ubiquitous principles of intelligence which are valid at each level of existence have been crystallised in the form of such simple glyphs. As said before in this book, glyphs are an extremely parsimonious way to store information. You may remember the set of mnemonic tools I developed in the form of glyphs, which I personally consider as a great "intuitive" tool to order my ideas and to spawn combinations thereof!

In this way, by tuning into this collective field of knowledge we are able to retrieve information going back aeons, which we then call intuition or creativity, but which in fact is merely a downloading of information from a mind at large.

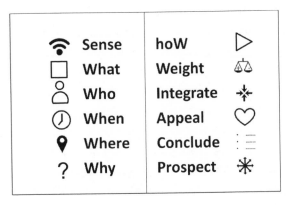

Figure 19: Glyph library

Other non-algorithmic activities

Noteworthy, intuition is not the only activity of the mind which does not appear to follow an algorithm. Various activities of the right hemisphere of the brain involving spatial orientation can – as of yet – not be described by algorithms. Such activities (e.g. rotation of objects) involve a mental visualisation, which takes place in the so-called "mind's eye". Neural network optimisation probably underlies such processes.

Conclusion

In this chapter I have speculated about the various ways intuition may arise. It is in the realm of intuition that intelligence follows a non-algorithmic method, which often gives better results than the best heuristic we can possibly think of. From quantum mechanics and digital physics, we have seen that reality might in fact be a kind of giant supercomputer, which is built from computational building blocks at its most basic level. This network combines principles of both neural networks and quantum computing principles and possibly has emergent effects going beyond these. Intuition may arise due to amplification of information though meta-system levels from a very basic level of primordial computation. Alternatively it may arise by tuning into and resonating with collective mind fields, which are ubiquitous in the animal kingdom and possibly exist at the human level as the collective unconscious.

We have also seen how the intelligence algorithm alternates order and chaos generating processes in a process of creating ever increasing levels of complexity. This is the way existence assures its self-sustention, its so-called "Autopoiesis". If reality is both a computer (but not as we know it) and a system with a global consciousness that integrates information and a global brain that spawns new combinations of information and if these processes take place at each and every level of existence, then perhaps reality is what we might call a process of "Pancomputational Panpsychism". But that's a story of one of my other books.[52]

This is my intuitive intelligence algorithm for you: Sense, Feel, Integrate, Spawn and Steem!

Appendix

Some elements of Chapter 1 as originally written.

Presently, we are living in the dawn of one of the greatest scientific breakthroughs: the very conceptualisation of the nature of intelligence, the self-organising pattern of the Universe.

In this appendix I will discuss the way Nature appears to follow a kind of algorithmic pattern of so-called "meta-system transitions" when it is applying intelligence in order to evolve.

In his book *Creating Internet Intelligence* Ben Goertzel is bringing the notions of "Complexity science" to a higher level of aggregation. Combining notions of Turchin's "metasystems transitions", Buddhism, General systems and Network theory and Peircean metaphysics, he tries to define the very essence of Intelligence. The insights presented in this book are of such a profound nature, that they may well one day be recognised as the ultimate intelligence algorithm that underlies every phenomenon in this universe.

A phrase that summarises the outcome of this algorithm is the "whole is more than the sum of its parts". Or put in one word: "Synergy" or "Emergence".

Let me summarise the deep philosophical background of this algorithm as presented in Chapter 2 of his book: "Elements of a Philosophy of Mind", where he starts with a summary of Peircean and Palmerian [hereinafter indicated in square brackets] metaphysics:

Naught is the original state of the universe or any other system. The formless void or undifferentiated state.

Firstness (raw being) is the conception (or cognitive realisation) of being or existing independent of anything else. This is what we philosophically could equate with "idealism". Its representation is a point. [Static, Being.]

Secondness (the reacting object) is the conception of being relative to, reaction with something else. This we could philosophically equate with "materialism". Its representation is a vector. [Dynamic, Becoming.] I'd also like to refer to this as the first step of the algorithm, which is a reaction to a stimulus resulting in a "polarisation".

Thirdness (evolving interpretation) is the conception of mediation, the step whereby first and second are brought in relation. Its representation is a triangle. [Philosophers refer to this as hyper emergent semi stasis, emerging form or dynamic/strange attractor.]

Then Goertzel adds a fourth element:

Fourthness (unity of consciousness) is the step of creating a pattern which emerges from a web of relationships which support and sustain each other so that the whole is greater than the sum of the parts. Its representation is a tetrahedron.

Goertzel has realised that this concept of emergence is the key of evolution. This is how a mind's intelligence comes into existence: The combination of two or more parts can lead to a new phenomenon, a new entity in which the whole is more than the sum of parts.

The new entity thus formed can be considered as a new firstness and can undergo this cycle again. When this compounded phenomenon interacts with another phenomenon, there is a new secondness, a new thirdness and fourthness (i.e. meta-tetrahedrons built of sub-tetrahedron building blocks). This is how complexity arises in every system. It is the core of evolution and intelligence. In the mind the ideas as vertices interact via the edges with other ideas associated with them. No idea has an independent existence but is compounded of features of other ideas and concepts. This creates a new idea by the virtue of emergence meaning that the whole is more than the sum of parts.

An element from Palmerian metaphysics,[2] which adds to these concepts, is the so-called "Wild Being", arising from the inter-action of the hyper emergent entities. This is the element of unexpected, unusual diversity generation which has an aspect of inspiration and intuition.

Let us for a moment leave this almost esoteric realm and return to the teachings of Ben Goertzel. Because there is more to the story of intelligence. With his previous companies Webmind, AGIRI, the Novamente project and his current program, the OpenCog project based on the work of volunteers, Goertzel et al have started defining what I would like to call "the laws of complex systems" and "the laws of Intelligence". Note that they do not claim to have achieved this; it is a tremendous task they have started, but it is all based on the law of emergence; meta-system transitions.

That the strong artificial intelligence promise of this approach has not been cashed in yet derives from the complexity of the system and physical constraints. As far as I understood it, these intelligent processes still take too much time in terms of *inter alia* response-time to be applicable in a "Global Brain" environment such as the Internet.

For intelligence to come to expression a substratum is necessary. In case of Goertzel's AI this can be the digital world of the Internet web.

Imagine that the intelligent system of Goertzel will one day be capable of expressing itself in a meaningful way in the Internet: will that system then arrive at self-awareness similar to ours?

The ingredients for an emergent consciousness appear to be there in terms of 1) knowledge per se; 2) the material/energetic substrate of the electronic environment; 3) the intelligent struc-tures of Goertzel as organising and hyper emergent structures, as a relation between 1 and 2. Could self-awareness emerge from this triad as the observer? Why not? In the end it is a form of consciousness at a different aggregation level born out of the consciousness of human beings.

When it comes to the tetrad knowledge-energy-intelligence-consciousness, one could even suppose that each one of these terms is a non-linear synergy of the three others. Given three of them you might inevitably evolve the fourth, or they are inextricable aspects of one and the same phenomenon. Existence as we know it could then be represented in a simplified form as the tetrahedron having these four concepts as knowable vertices and functioning as the creative wholeness which is the ultimate reality.

What a pity, that a company as "Webmind" was simply lost due to bankruptcy as a consequence of a lack of sufficient investors.

The principles of chapter 1 can be applied to self-learning AI systems.

It is to be noted that this seven-step algorithm represents the inventive or creative exponent of intelligence. What distinguishes true intelligence the most from savantism is its focus when searching a successful strategy, without getting lost in irrelevant details. The search for successful strategies and the storing of corresponding so-called "heuristics" essentially involve the steps 1–3 of the above given algorithm. If a strategy is sufficiently similar to an existing successful one, there is no need for inventive recombination and this very strategy will be selected, thus leaving out steps 4–7. A need for inventive creativity arises when the system is under resource-limitation stress.

This is the intelligence algorithm: Probing a diversity generating antithesis as a result of a stimulus from the inadequacy of the status quo thesis (e.g. a lack of resources), pattern abstraction, emergence of multiple alternative strategies, intergroup tournaments and distinction probing resulting in either niching or preferably symbiosis.

The most promising strategies ideally result in symbiosis, a unification of features toward which the system will strive. It will try to morphogenetically resonate with its new environment and thereby adapt to it.

In the case of artificial intelligence (AI) *casu quo* a "webmind", e.g. a neural network based Global Brain such as the Internet, it can be said that if the system is put under pressure due to scarcity of resources, it is indispensable; it has a way to venture into the unknown to discover new resources.

Yet the system as a whole cannot venture into the unknown by making a big leap; that is simply too risky. A webmind apparently disposing of free will is therefore ideally a Society-of-Minds, wherein the different individuals have been attributed the roles of conformity enforcers, inner judges, resource shifters and diversity generators, so that the system as a whole can safely sacrifice diversity generators on a massive scale, without compromising the integrity of the whole in order to find new promising strategies, heuristics and/or resources.

Among the diversity generators as algorithm-generating aLife AI agents it can be envisaged that there are different groups or ensembles each having a different degree of freedom to explore: There can be a gradual increase from rather conservative combinations of existing strategies that a diversity generator can propose, until absurd wild combinations of unrelated strategies. Conservative diversity generators will still look for certain degrees of resemblances between existing strategies (in the form of algorithms) and combine parts of these linearly. When more freedom is allowed non-linear combinations can be used and the most free systems can have access to random combinations on the verge of the absurd. Imagine these as algorithm modifying aLife AI agents, which build a combinatorial library of an algorithm-type Lego and start multiple rounds of screening for a given desired result. The diversity generators themselves are still algorithm bound, but a successful one will be seen by the outside world as having had a great deal of free will.

Evolution of colony-based organisms and cell aggregates

within an organism works in a similar way: Think of the hyper-mutation process of the immune system and the recently developed phage assisted directed evolution.

The big advantage of a future AI based hypermutation as an in silico equivalent of in vitro "Directed evolution" is that it is faster than both traditional evolution and intelligent design by humans. Whereas intelligent design is limited by what the system knows and whereas traditional evolution is limited by the resource limitation parameters of the environment, directed evolution can perform massively parallel screening for a given characteristic and select a candidate fulfilling those require-ments. That candidate itself becomes the new scaffold (and emergent entity) to modify, like the further modification of selected lead compounds from combinatorial chemistry libraries, which have been identified to show the desired activity. Based on that scaffold a new round of building a library and repeating the screening and selection process can be carried out. Multiple rounds of selection lead to hypermutation, similarly to what happens in the immune system. The outcome of the final extremely high affinity product cannot be predicted a priori and will be considered by an observer from the outside as an utterly original decision deriving from what we would identify as an act out of free will.

This can for instance be applied in the evolutionary design of "Motome" elements for robots linked to the Webmind. (The Motome is the complete collection of entities endowed with motricity linked to the network.) If several Motome solutions (tendons, wheels, tracks) are known then combinatorial libraries of these (inter-species) can be probed in different virtual environ-ments. In addition the individual building blocks themselves (tendons, wheels, tracks) can be evolved at a different aggre-gation level (intra-species) leading to new Lego blocks adapted to a given environment and avoiding a quantisation problem.

Similar random combination events already occur in robot

ensembles generating a robot language: The so-called "Lingodroids".[53]

If this algorithm-Lego protocol is carried out on a material level with different types of programmed nanites capable of self-assembly, Vernor Vinge's[54] morphing ideas can become a reality: A nanite embodied AI entity can then morphogenetically adapt to its environment and take any form needed: It is the carbon-based evolution as we know it, repeated in silico at a much higher speed. Moreover the individual species themselves become morphogenetically alterable chameleons depending on the environment they encounter. Each individual can take any shape and saves the "phylum"-patterns that are appropriate for a given environment as part of its learning abilities. (A phylum is a biological taxonomic rank between "kingdom" and "class" and represents a specific morphological pattern.)

What we call utterly original inventive intelligence occurs when the problem-solving features of a solution in an analogous problem situation (from a relatively distant other domain or technical field) are applied to an existing entity or process to solve a problem. There is then a unilateral exchange or rather an addition of features. Screening swarms of algorithm modifying bots that seek solutions in distant fields may not have the highest chance of success, but if they do they may cause breakthroughs that liberate a system under resource pressure, where conservative attempts to solutions would have failed.

My thesis here is not about the complete set of ingredients for artificial general intelligence as a whole, but the inventive exponents thereof. To apply solutions from distant fields is non-obvious. The ones familiar with patent-drafting will recognise this. Another form of apparent inventive skill derives from serendipity, when upon searching for a solution to a given problem one stumbles upon a solution to a different problem.

This is by the way different from the type of blissful realisations that a solution for a given problem can advantageously be applied for a different purpose, leading in fact to the problem being defined after the solution having been envisaged: The so-called "problem-inventions".

When Dietrich Dörner's[30] AI "Steamcar" ("Kesselwagen" in the book *Bauplan für eine Seele*) encountered a resource restriction, i.e. no water was available on ground level, it used trial and error behaviour with its existing tools: thus its proboscis to suck water was applied for a different purpose, namely to hit a tree, which made water accumulated in the leaves fall on the ground, thereby resolving the resource problem of the entity. Via its "quasineuronal" (i.e. neural network) structure involving a need-indicator, it was able to learn a new solution for its goal. The need transformed into a "motif", which is in fact a need plus a goal-indication.

In human beings the utterly inventive connections in the brains are provided by the so-called spindle cells, which wire up totally unrelated areas of the brain. This is also a feature which discriminates us from most other higher mammals.

We as human beings may also fulfil the roles of the different types of individuals of a Society-of-Minds. The universe is probing for new solutions in order to propagate its seven-step intelligence algorithm and it also uses us to achieve that goal.

References

1. Antonin Tuynman. *Technovedanta, Internet Architecture of a Quasiconscious Vedantic Webmind: A Panpsychic Theory of Everything*, Lulu, 2012.
2. Ben Goertzel. *Creating Internet Intelligence: Wild Computing, Distributed Digital Consciousness, and the Emerging Global Brain, IFSR International Series on Systems Science and Engineering, Vol. 18*, Kluwer Academic/Plenum Publishers, 2002.
3. RR Llinás. *I of the Vortex: From Neurons to Self*, MIT Press, 2002.
4. Eshel Ben-Jacob. "Bacterial wisdom, Gödel's theorem and creative genomic webs", *Physica A*, 248, pp. 57–76, 1998. Note that the terminology "conformity enforcers, diversity generators, resource shifters, inner judges and intergroup tournament" is a terminology introduced by Howard Bloom, based on Ben-Jacob's theory.
5. Howard Bloom. *Global Brain: The Evolution of Mass Mind from the Big Bang to the 21st Century*, Wiley, 2000.
6. Terence McKenna. *Food of the Gods: The Search for the Original Tree of Knowledge*, Rider & Co, 1999.
7. R. Buckminster Fuller. *Synergetics: Explorations in the Geometry of Thinking*, Macmillan, 1982.
8. JF Herbart. "Definition of Apperception" [Online]. Available from: https://en.wikipedia.org/wiki/Apperception
9. Verlinde, EP. "On the Origin of Gravity and the Laws of Newton", *JHEP*, 1104:029, 2011.
10. Patanjali. *Patanjali's Yoga Sutra*, Penguin Classics, 2009.
11. Nagarjuna. *The Fundamental Wisdom of the Middle Way: Nagarjuna's Mulamadhyamakakarika*, Oxford University Press, 1995.
12. RA Wilson. *Prometheus Rising*, New Falcon Publications,

1983.

13. Bernardo Kastrup. *Meaning in Absurdity*, John Hunt Publishing, 2012.

14. G. Pólya. *How to Solve It*, Penguin Books Ltd, 1990.

15. Kuhn, HW; Nasar, S. *The Essential John Nash*, Princeton, 2002.

16. William Marston. *Emotions of Normal People*, Cooper Press, 2014.

17. Maslow, AH. "A theory of human motivation", *Psychological Review*, 50 (4): 370–396, 1943.

18. Robert Plutchik. *Emotions and Life: Perspectives from Psychology, Biology, and Evolution*, American Psychological Association, 2002.

19. Eric Berne. *Transactional Analysis in Psychotherapy*, Martino Fine Books, 2015.

20. John 8:7, the Bible, New Testament.

21. Kim Eng. 2004 [Online]. Available from: https://www.eckharttolle.com/article/Relationships-True-Love-and-the-Transcendence-of-Duality

22. Sadhguru Jaggi Vasudev. *Inner Engineering: A Yogi's Guide to Joy*, Spiegel & Grau, 2016.

23. Thich Nhat Hanh. *Old Path White Clouds: Walking in the Footsteps of the Buddha*, Parallax Press, 1991.

24. Arthur C. Clarke. *2001: A Space Odyssey*, Roc, Reissue edition, 2000.

25. Moebius. "Red-Beard and the Brain Pirate", *Heavy Metal*, v. 4, no. 8, pp. 79–83, 1980.

26. Tim Gross. 2011 [Online]. Available from: http://www.kurzweilai.net/forums/topic/network-dynamics-the-secret-of-all-processes-in-the-cosmos#post-18431

27. Oizumi, M; Albantakis, L; Tononi, G. "From the Phenomenology to the Mechanisms of Consciousness: Integrated Information Theory 3.0". *PLoS Comput Biol*, 10(5): e1003588, 2014.

28. Isaac Asimov. *I, Robot*, Gnome Press, 1950.

29. Martin Lodewijk. *The Pirates of Pandarve*, Big Balloon Publishers, 1983.

30. Dietrich Dörner. *Bauplan für eine Seele*, Rowohlt Taschenbuch Verlag, 2001.

31. Howard Bloom. *The Genius of the Beast*, Prometheus Books, 2010.

32. Ben Goertzel. *The Hidden Pattern*, Brown Walker Press, 2006.

33. John Searle. "Mind, brains, and programs", *Behavioral and Brain Sciences*, 3, pp. 417–424, 1980.

34. J-P Vasseur and Adam Dunkels. *Interconnecting Smart Objects with IP: The Next Internet*, Morgan Kaufmann, 2010.

35. Kurzweil, R. *How to Create a Mind*, Viking Penguin, 2012.

36. MN Huhns. "The Sentient Web", *IEEE Internet Computing*, issue Nov-Dec., pp. 82–84, 2003.

37. Yves Couder. "Silicone Oil Droplets showing quantum like interference", 2011 [Online]. Available from: https://www.youtube.com/watch?v=GHHaDWEWtQE

38. Kim, Yoon-Ho; R. Yu; SP Kulik; YH Shih; Marlan Scully. "A Delayed 'Choice' Quantum Eraser". *Physical Review Letters*, 84: 1–5, (2000).

39. Dean Radin. *Supernormal*, Deepak Chopra Books, 2013.

40. Ervin Laszlo. *Science and the Akashic Field*, Inner Traditions, 2004.

41. Ralph Abraham and Sisir Roy. *Demystifying the Akasha*, Epigraph Books, 2010.

42. Kun Xie, Grace E. Fox, Jun Liu, Cheng Lyu, Jason C. Lee, Hui Kuang, Stephanie Jacobs, Meng Li, Tianming Liu, Sen Song, Joe Z. Tsien. "Brain Computation Is Organized via Power-of-Two-Based Permutation Logic", *Frontiers in Systems Neuroscience*, 2016.

43. Penrose, R. *The Emperor's New Mind*, Vintage, 1990.

44. SJ Gates, et al. "Relating Doubly-Even Error-Correcting Codes, Graphs, and Irreducible Representations of N-Extended Supersymmetry". 2008 [Online]. Available from:

http://arxiv.org/abs/0806.0051

45. Wheeler, John A. "Information, physics, quantum: The search for links". In Zurek, Wojciech Hubert, *Complexity, Entropy and the Physics of Information*, Redwood City, California: Addison-Wesley, 1990.

46. Bostrom, N. "Are You Living in a Simulation?", *Philosophical Quarterly*, Vol. 53, No. 211, pp. 243–255, 2003.

47. DR Hofstadter. *Gödel, Escher, Bach: an Eternal Golden Braid*, Penguin Books, 1979.

48. S. Kaufman. *Unified Reality Theory: The Evolution of Existence Into Experience*, Destiny Toad Press, 2002.

49. Johannes Koelman. 2010 [Online]. Available from: http://www.science20.com/hammock_physicist/it_bit_entropic_gravity_pedestrians-66244

50. James Gleick. *Chaos: Making a New Science*, Viking, 1987.

51. Rupert Sheldrake. *The Presence of the Past*, Collins, 1988.

52. A. Tuynman. *Technovedanta 2.0: Transcendental Metaphysics of Pancomputational Panpsychism*, Lulu, 2016.

53. Evan Ackerman. "Lingodroid Robots Invent Their Own Spoken Language", *IEEE Spectrum*, 2011 [Online]. Available from http://spectrum.ieee.org/automaton/robotics/artificial-intelligence/lingodroid-robots-invent-their-own-spoken-language

54. Vernor Vinge. 1993, "Vernor Vinge on the Singularity" [Online]. Available from: http://mindstalk.net/vinge/vinge-sing.html

Recent bestsellers from Iff Books are:

Why Materialism Is Baloney
How True Skeptics Know There is no Death and Fathom
Answers to Life, the Universe, and Everything
Bernardo Kastrup
A hard-nosed, logical, and skeptic non-materialist metaphysics,
according to which the body is in mind, not mind in the body.
Paperback: 978-1-78279-362-5 ebook: 978-1-78279-361-8

The Fall
Steve Taylor
The Fall discusses human achievement versus the issues of war,
patriarchy and social inequality.
Paperback: 978-1-90504-720-8 ebook: 978-184694-633-2

Brief Peeks Beyond
Critical Essays on Metaphysics, Neuroscience, Free Will,
Skepticism and Culture
Bernardo Kastrup
An incisive, original, compelling alternative to current
mainstream cultural views and assumptions.
Paperback: 978-1-78535-018-4 ebook: 978-1-78535-019-1

Framespotting
Changing How You Look at Things Changes How
You See Them
Laurence & Alison Matthews
A punchy, upbeat guide to framespotting. Spot deceptions and
hidden assumptions; swap growth for growing up. See and be
free.
Paperback: 978-1-78279-689-3 ebook: 978-1-78279-822-4

Is There an Afterlife?
David Fontana

Is there an Afterlife? If so what is it like? How do Western ideas of the afterlife compare with Eastern? David Fontana presents the historical and contemporary evidence for survival of physical death.
Paperback: 978-1-90381-690-5

Nothing Matters
A Book About Nothing
Ronald Green

Thinking about Nothing opens the world to everything by illuminating new angles to old problems and stimulating new ways of thinking.
Paperback: 978-1-84694-707-0 ebook: 978-1-78099-016-3

Panpsychism
The Philosophy of the Sensuous Cosmos
Peter Ells

Are free will and mind chimeras? This book, anti-materialistic but respecting science, answers: No! Mind is foundational to all existence.
Paperback: 978-1-84694-505-2 ebook: 978-1-78099-018-7

Punk Science
Inside the Mind of God
Manjir Samanta-Laughton

Many have experienced unexplainable phenomena; God, psychic abilities, extraordinary healing and angelic encounters. Can cutting-edge science actually explain phenomena previously thought of as 'paranormal'?
Paperback: 978-1-90504-793-2

The Vagabond Spirit of Poetry
Edward Clarke
Spend time with the wisest poets of the modern age and of the past, and let Edward Clarke remind you of the importance of poetry in our industrialized world.
Paperback: 978-1-78279-370-0 ebook: 978-1-78279-369-4

Readers of ebooks can buy or view any of these bestsellers by clicking on the live link in the title. Most titles are published in paperback and as an ebook. Paperbacks are available in traditional bookshops. Both print and ebook formats are available online.

Find more titles and sign up to our readers' newsletter at http://www.johnhuntpublishing.com/non-fiction

Follow us on Facebook at https://www.facebook.com/JHPNonFiction and Twitter at https://twitter.com/JHPNonFiction